SEDIMENTOLOGY AND STRATIGRAPHY

Edited by
Judith Rosales

Delve Publishing

Sedimentology and Stratigraphy

Edited by Judith Rosales

Edition 2018

Delve Publishing

Published by
Delve Publishing LLC
708 3rd Avenue, 6th Floor
New York
NY 10801
United States of America
www.delvepublishing.com

ISBN: 978-1-68095-667-2 (Hardback)
Printed in United States of America

Library of Congress Control Number: 2016920418

About the Editor

Judith Rosales

Prof. Judith is an Ecologist and Environmental Specialist with more than 30 years of academic and consulting experience. She holds a PhD in geography from the University of Birmingham UK and wrote the PhD and Master Academic Programs of Environmental Sciences for the Universidad de Guayana, Venezuela.

Table of Contents

3 SEDIMENTARY ROCKS

4 SEDIMENT TRANSPORT

5 FLUVIAL DEPOSITS

6 GLACIAL ENVIRONMENTS

7 ALLUVIUM

8 STRATIGRAPHY

List of Abbreviations

ESR	Especially Siliciclastic Rocks
ECR	Especially Carbonate Rocks
GMPTS	Global Magnetic Polarity Time Scale
PES	Particle Erosion Simulator
SAR	Synthetic Aperture Radar
SSD	Soft Sediment Deformation

Preface

Sedimentology studies sand, mud (or silt) and clay, and the various ways they are deposited. Using these studies, sedimentologists apply their understanding of modern processes to ancient rock, to try to understand how it formed. Most of the rocks on earth are sedimentary rocks, and it is in these kinds of rocks you find fossils and many of the other historical markers. Sedimentary rocks are also where petroleum deposits are found. Sedimentology is tied to stratigraphy, which studies the relationships between rock layers and how they can shift and move. This also affects where petroleum deposits can be found, as well as how the extraction of petroleum affects the sediment around the deposit. Sedimentary rocks cover up to 75% of the Earth's surface, record much of the Earth's history, and harbor the fossil record. Sedimentology is closely linked to stratigraphy, the study of the physical and temporal relationships between rock layers or strata. The premise that the processes affecting the earth today are the same as in the past is the basis for determining how sedimentary features in the rock record were formed. The aim of sedimentology, studying sediments, is to derive information on the depositional conditions which acted to deposit the rock unit, and the relation of the individual rock units in a basin into a coherent understanding of the evolution of the sedimentary sequences and basins, and thus, the Earth's geological history as a whole.

The scientific basis of this is the principle of uniformitarianism, which states that the sediments within ancient sedimentary rocks were deposited in the same way as sediments which are being deposited at the Earth's surface today. This Text introduces the reader to sedimentology and stratigraphic principles, and provides tools for the interpretation of sediments and sedimentary rocks.

Content Coverage

Chapter One presents the introduction of sedimentation, the process of accumulating sediments, or dirt, on the reef. It is a natural phenomenon that results from land and reef erosion.

Chapter Two focuses on sedimentary structures, which are visible features within sedimentary rocks that formed at the time of deposition and represent manifestations of the physical and biological processes that operated in depositional environments.

Chapter Three focuses on sedimentary rocks that are formed by the deposition and subsequent cementation of that material at the Earth's surface and within bodies of water.

Chapter Four highlights about sediment transport occurs in natural systems where the particles are clastic rocks (sand, gravel, boulders, etc.), mud, or clay; the fluid is air, water, or ice; and the force of gravity acts to move the particles along the sloping surface on which they are resting.

Chapter Five focuses on fluvial processes that include the motion of sediment and erosion or deposition on the river bed.

Chapter Six emphasizes on glacial environments that represent very harsh but beautiful environments and are located in the Polar Regions or in high mountainous regions.

Chapter Seven focuses on alluvium that can contain valuable ores such as gold and platinum and a wide variety of gemstones. Such a concentration of valuable ores is termed a placer deposit.

Chapter Eight presents an overview of stratigraphy, which studies rock layers (strata) and layering (stratification). It is primarily used in the study of sedimentary and layered volcanic rocks.

CHAPTER 1

Sediment and Sedimentation

Chapter Contents

- **Introduction**
- **Classification**
- **Shores and Shallow Seas**
- **Environmental Issues**
- **Sedimentation**
- **Sedimentation Equilibrium**

In geological sciences, sedimentation is the process of accumulating sediments, or dirt, on the reef. It is a natural phenomenon that results from land and reef erosion. Broadly defined, it also encompasses glacial ice deposits and accumulation of materials due to gravity, such as talus deposits and rock debris collection at the base of cliffs.

Soil decomposition, plant decomposition, animal decomposition and bedrock erosion are natural sources of sediments transported to the oceans. Sediments often add nutrients to the soil. Natural sediment mobilization helps develop and maintain coastal habitats, such as mangroves, coral reefs, estuaries, lagoons and wetlands. However, many human activities often alter the process of sedimentation and erosion. High sedimentation destroys corals when sediments land directly on top of corals, smothering the corals. Corals are not capable of removing sediments, because they cannot move. Sedimentation is sometimes detrimental to corals, mainly because it affects water clarity and makes the water cloudy or turbid. When there is high turbidity, the amount of light that reaches the corals is considerably decreased. With lesser light, corals find it difficult to photosynthesize. Human activities are major culprits in the increase of sedimentation rates along many coastline areas. These activities are mainly related to poor land management practices, such as unregulated use of off-road vehicles and poor urban development.

INTRODUCTION

Sediment is a naturally occurring material that is broken down by processes of weathering and erosion, and is subsequently transported by the action of wind, water, or ice, and/or by the force of gravity acting on the particles. For example, sand and silt can be carried in suspension in river water and on reaching the sea be deposited by sedimentation and if buried this may eventually become sandstone and siltstone, (sedimentary rocks).

Sediments are most often transported by water (fluvial processes), but also wind (aeolian processes) and glaciers. Beach sands and river channel deposits are examples of fluvial transport and deposition, though sediment also often settles out of slow-moving or standing water in lakes and oceans. Desert sand dunes and loess are examples of aeolian transport and deposition. Glacial moraine deposits and till are ice-transported sediments.

Cobbles on a beach.

River Rhône flowing into Lake Geneva. Sediments make the water appear brownish-grey; they are an indicator of increased water runoff, land degradation, erosion due to intensive industrialized land use, land sealing, and poor soil management.

Sediment billowing out from Italy's shore into the Adriatic Sea.

CLASSIFICATION

Sediment can be classified based on its grain size and/or its composition.

Grain Size

Particle size, also called grain size, refers to the diameter of individual grains of sediment, or the lithified particles in clastic rocks. The term may also be applied to other granular materials. This is different from the crystallite size, which refers to the size of a single crystal inside a particle or grain. A single grain can be composed of several crystals. Granular material can range from very small colloidal particles, through clay, silt, sand, gravel, and cobbles, to boulders.

Krumbein Phi Scale

Size ranges define limits of classes that are given names in the Wentworth scale (or Udden–Wentworth scale) used in the United States. The Krumbein *phi* (φ) scale, a modification of the Wentworth scale created by W. C. Krumbein in 1937, is a logarithmic scale computed by the equation

$$\phi = -\log_2 D/D_0,$$

where

ϕ is the Krumbein phi scale,

D is the diameter of the particle or grain in millimeters (from petrowiki, krumbein and monks equation) and

D_0 is a reference diameter, equal to 1 mm (to make the equation dimensionally consistent).

This equation can be rearranged to find diameter using φ:

$$D = D_0 \times 2^{-\phi}$$

Sediment in the Gulf of Mexico.

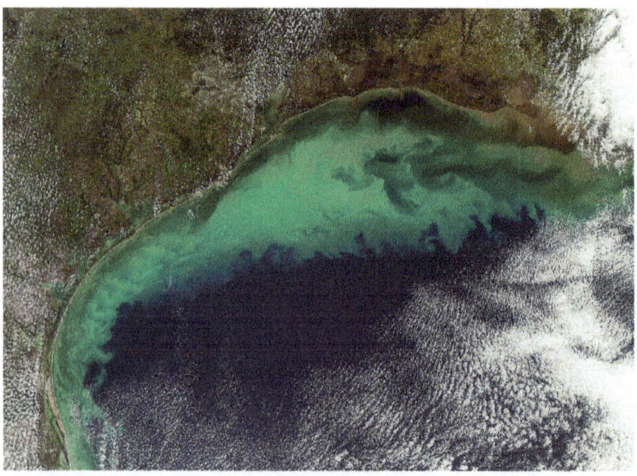

Sediment off the Yucatán Peninsula.

Sediment size is measured on a log base 2 scale, called the "Phi" scale, which classifies particles by size from "colloid" to "boulder".

φ scale	Size range (metric)	Size range (inches)	Aggregate class (Wentworth)	Other names
< -8	> 256 mm	> 10.1 in	Boulder	
-6 to -8	64–256 mm	2.5–10.1 in	Cobble	
-5 to -6	32–64 mm	1.26–2.5 in	Very coarse gravel	Pebble
-4 to -5	16–32 mm	0.63–1.26 in	Coarse gravel	Pebble
-3 to -4	8–16 mm	0.31–0.63 in	Medium gravel	Pebble
-2 to -3	4–8 mm	0.157–0.31 in	Fine gravel	Pebble
-1 to -2	2–4 mm	0.079–0.157 in	Very fine gravel	Granule
0 to -1	1–2 mm	0.039–0.079 in	Very coarse sand	
1 to 0	0.5–1 mm	0.020–0.039 in	Coarse sand	
2 to 1	0.25–0.5 mm	0.010–0.020 in	Medium sand	
3 to 2	125–250 μm	0.0049–0.010 in	Fine sand	
4 to 3	62.5–125 μm	0.0025–0.0049 in	Very fine sand	
8 to 4	3.9–62.5 μm	0.00015–0.0025 in	Silt	Mud
> 8	< 3.9 μm	< 0.00015 in	Clay	Mud
>10	< 1 μm	< 0.000039 in	Colloid	Mud

International Scale

ISO 14688-1:2002, establishes the basic principles for the identification and classification of soils on the basis of those material and mass characteristics most commonly used for soils for engineering purposes. ISO 14688-1 is applicable to natural soils *in situ*, similar man-made materials *in situ* and soils redeposited by people.

ISO 14688-1:2002					
Name				**Size range (mm)**	**Size range (approx. in)**
Very coarse soil		Large boulder	LBo	>630	>24.8031
Boulder		Bo	200–630	7.8740–24.803	
Cobble		Co	6 3 – 200	2.4803–7.8740	
Coarse soil	Gravel	Coarse gravel	CGr	20–63	0.78740–2.4803
		Medium gravel	MGr	6.3–20	0.24803–0.78740
		Fine gravel	FGr	2.0–6.3	0.078740–0.24803
	Sand	Coarse sand	CSa	0.63–2.0	0.024803–0.078740
		Medium sand	MSa	0.2–0.63	0.0078740–0.024803
		Fine sand	FSa	0.063–0.2	0.0024803–0.0078740
Fine soil	Silt	Coarse silt	CSi	0.02–0.063	0.00078740–0.0024803
		Medium silt	MSi	0.0063–0.02	0.00024803–0.00078740
		Fine silt	FSi	0.002–0.0063	0.000078740–0.00024803
	Clay		Cl	≤0.002	≤0.000078740

Sorting

An accumulation of sediment can also be characterized by the grain size distribution. A sediment deposit can undergo sorting when a particle size range is removed by an agency such as a river or the wind. The sorting can be quantified using the Inclusive Graphic Standard Deviation:

$$\sigma_I = \frac{\phi 84 - \phi 16}{4} + \frac{\phi 95 - \phi 5}{6.6}$$

where

σ_I is the Inclusive Graphic Standard Deviation in phi units

$\phi 84$ is the 84th percentile of the grain size distribution in phi units, etc.

The result of this can be described using the following terms:

$\sigma_I < 0.35$	$0.35 < \sigma_I < 0.50$	$0.50 < \sigma_I < 0.71$	$0.71 < \sigma_I < 1.00$	$1.00 < \sigma_I < 2.00$	$2.00 < \sigma_I$
very well sorted	well sorted	moderately well sorted	moderately sorted	poorly sorted	very poorly sorted

Composition

Composition of sediment can be measured in terms of:

- parent rock lithology
- mineral composition
- chemical make-up.

This leads to an ambiguity in which clay can be used as both a size-range and a composition.

Sediment Transport

Sediment builds up on human-made breakwaters because they reduce the speed of water flow, so the stream cannot carry as much sediment load.

Glacial transport of boulders. These boulders will be deposited as the glacier retreats.

Sediment is transported based on the strength of the flow that carries it and its own size, volume, density, and shape. Stronger flows will increase the lift and drag on the particle, causing it to rise, while larger or denser particles will be more likely to fall through the flow.

Rouse Number

The Rouse number (P or Z) is a non-dimensional number in fluid dynamics which is used to define a concentration profile of suspended sediment and which also determines how sediment will be transported in a flowing fluid. It is a ratio between the sediment fall velocity w_s and the upwards velocity on the grain as a product of the von Kármán constant κ and the shear velocity u_*.

$$\mathrm{P} = \frac{w_s}{\kappa u_*}$$

Occasionally the factor β is included before the von Kármán constant in the equation, which is a constant which correlates eddy viscosity to eddy diffusivity. This is generally taken to be equal to 1, and therefore is ignored in actual calculation. However, it should not be ignored when considering the full equation.

$$\mathrm{P} = \frac{w_s}{\beta \kappa u_*}$$

It is named after the American fluid dynamicist Hunter Rouse. It is a characteristic scale parameter in the Rouse Profile of suspended sediment concentration with depth in a flowing fluid. The concentration of suspended sediment with depth goes as the power of the negative Rouse number. It also is used to determine how the particles will move in the fluid. The required Rouse numbers for transport as bed load, suspended load, and wash load, are given below.

Mode of Transport	Rouse Number
Bed load	>2.5
Suspended load: 50% Suspended	>1.2, <2.5
Suspended load: 100% Suspended	>0.8, <1.2
Wash load	<0.8

Aeolian Processes: Wind

Wind results in the transportation of fine sediment and the formation of sand dune fields and soils from airborne dust.

Glacial Processes

Glacial sediments from Montana.

Glaciers carry a wide range of sediment sizes, and deposit it in moraines.

ENVIRONMENTAL ISSUES

Erosion and Agricultural Sediment Delivery to Rivers

One cause of high sediment loads from slash and burn and shifting cultivation of tropical forests. When the ground surface is stripped of vegetation and then seared of all living organisms, the upper soils are vulnerable to both wind and water erosion. In a number of regions of the earth, entire sectors of a country have become erodible. For example, on the Madagascar high central plateau, which constitutes approximately ten percent of that country's land area, most of the land area is devegetated, and gullies have eroded into the underlying soil in furrows typically in excess of 50 meters deep and one kilometer wide. This siltation results in discoloration of rivers to a dark red brown color and leads to fish kills.

Erosion is also an issue in areas of modern farming, where the removal of native vegetation for the cultivation and harvesting of a single type of crop has left the soil unsupported. Many of these regions are near rivers and drainages. Loss of soil due to erosion removes useful farmland, adds to sediment loads, and can help transport anthropogenic fertilizers into the river system, which leads to eutrophication.

Coastal Development and Sedimentation near Coral Reefs

Watershed development near coral reefs is a primary cause of sediment-related coral stress. The stripping of natural vegetation in the watershed for development exposes soil to increased wind and rainfall, and as a result, could cause exposed sediment to become more susceptible to erosion and delivery to the marine environment during rainfall events. Sediment can negatively affect corals in many ways, such as by physically smothering them, abrading their surfaces, causing corals to expend energy during sediment removal, and causing algal blooms that can ultimately lead to less space on the seafloor where juvenile corals (polyps) can settle.

When sediments are introduced into the coastal regions of the ocean, the proportion of land, marine and organic-derived sediment that characterizes the seafloor near sources of sediment output is altered. In addition, because the source of sediment (i.e. land, ocean, or organically) is often correlated with how coarse or fine sediment grain sizes that characterize an area are on average, grain size distribution of sediment will shift according to relative input of land (typically fine), marine (typically coarse), and organically-derived (variable with age) sediment. These alterations in marine sediment characterize the amount of sediment that is suspended in the water column at any given time and sediment-related coral stress.

SEDIMENTATION

Sedimentation is the tendency for particles in suspension to settle out of the fluid in which they are entrained and come to rest against a barrier. This is due to their motion through the fluid in response to the forces acting on them: these forces can be due to gravity, centrifugal acceleration, or electromagnetism. In geology, sedimentation is often used as the opposite of erosion, i.e., the terminal end of sediment transport. In that sense, it includes the termination of transport by saltation or true bedload transport. Settling is the falling of suspended particles through the liquid, whereas sedimentation is the termination of the settling process.

Sedimentation may pertain to objects of various sizes, ranging from large rocks in flowing water to suspensions of dust and pollen particles to cellular suspensions to solutions of single molecules such as proteins and peptides. Even small molecules supply a sufficiently strong force to produce significant sedimentation.

The term is typically used in geology to describe the deposition of sediment which results in the formation of sedimentary rock, but it is also used in various chemical and environmental fields to describe the motion of often-smaller particles and molecules. This process is also used in the biotech industry to separate cells from the culture media.

Experiments

In a sedimentation experiment, the applied force accelerates the particles to a terminal velocity v_{term} at which the applied force is exactly canceled by an opposing drag force. For small enough particles (low Reynolds number), the drag force varies linearly with the terminal velocity, i.e., $F_{drag} = fv_{term}$ (Stokes flow) where f depends only on the properties of the particle and the surrounding fluid. Similarly, the applied force generally varies linearly with some coupling constant (denoted here as q) that depends only on the properties of the particle, $F_{\text{app}} = qE_{\text{app}}$. Hence, it is generally possible to define a sedimentation coefficient $s \stackrel{\text{def}}{=} q/f$ that depends only on the properties of the particle and the surrounding fluid. Thus, measuring s can reveal underlying properties of the particle.

In many cases, the motion of the particles is blocked by a hard boundary; the resulting accumulation of particles at the boundary is called a sediment. The concentration of particles at the boundary is opposed by the diffusion of the particles.

The sedimentation of a single particle under gravity is described by the Mason–Weaver equation, which has a simple exact solution. The sedimentation coefficient s in this case equals m_b/f, where m_b is the buoyant mass.

The sedimentation of a single particle under centrifugal force is described by the Lamm equation, which likewise has an exact solution. The sedimentation coefficient s also equals m_b/f, where m_b is the buoyant mass. However, the Lamm equation differs from the Mason–Weaver equation because the centrifugal force depends on radius from the origin of rotation, whereas in the Mason–Weaver equation gravity is constant. The Lamm equation also has extra terms, since it pertains to sector-shaped cells, whereas the Mason–Weaver equation is one-dimensional.

Classification of sedimentation:

- Type 1 sedimentation is characterized by particles that settle discretely at a constant settling velocity,or by the deposition of Iron-Rich minerals to streamlines down to the point source. They settle as individual particles and do not flocculate or stick to other during settling. Example: sand and grit material
- Type 2 sedimentation is characterized by particles that flocculate during sedimentation and because of this their size is constantly changing and therefore their settling velocity is changing. Example: alum or iron coagulation
- Type 3 sedimentation is also known as zone sedimentation. In this process the particles are at a high concentration (greater than 1000 mg/L) such that the particles tend to settle as a mass and a distinct clear zone and sludge zone are present. Zone settling occurs in lime-softening, sedimentation, active sludge sedimentation and sludge thickeners.

Geology

In geology, sedimentation is the deposition of particles carried by a fluid flow. For suspended load, this can be expressed mathematically by the Exner equation, and results in the formation of depositional landforms and the rocks that constitute sedimentary record. An undesired increased transport and sedimentation of suspended material is called siltation, and it is a major source of pollution in waterways in some parts of the world. Climate change also affects siltation rates.

Siltation.

Chemistry

In chemistry, sedimentation has been used to measure the size of large molecules (macromolecule), where the force of gravity is augmented with centrifugal force in an ultracentrifuge.

SEDIMENTATION EQUILIBRIUM

Sedimentation equilibrium in a solution or suspension of different particles, such as molecules, exists when the rate of transport of each material in any one direction due to sedimentation equals the rate of transport in the opposite direction due to diffusion. Sedimentation is due to an external force, such as gravity (for very large particles) or centrifugal force in a centrifuge.

Modern applications use the analytical ultracentrifuge. The theoretical basis for the measurements is developed from the Mason-Weaver equation. The advantage of using analytical sedimentation equilibrium analysis for Molecular Weight of proteins and their interacting mixtures is the avoidance of need for derivation of a frictional coefficient, otherwise required for interpretation of dynamic sedimentation.

It was discovered for large particles by Jean Baptiste Perrin for which he received the Nobel Prize in Physics in 1926.

Application in Determining Molecular Mass

Sedimentation equilibrium can be used to determine molecular mass. It forms the basis for an analytical ultracentrifugation method for measuring molecular masses, such as those of proteins, in solution.

REFERENCES

1. Prothero, Donald R.; Schwab, Fred (1996), Sedimentary Geology: An Introduction to Sedimentary Rocks and Stratigraphy, W. H. Freeman, ISBN 0-7167-2726-9
2. Siever, Raymond (1988), Sand, New York: Scientific American Library, ISBN 0-7167-5021-X
3. Nichols, Gary (1999), Sedimentology & Stratigraphy, Malden, MA: Wiley-Blackwell, ISBN 0-632-03578-1
4. Reading, H. G. (1978), Sedimentary Environments: Processes, Facies and Stratigraphy, Cambridge, MA: Blackwell Science, ISBN 0-632-03627-3
5. Krumbein, W. C.; Aberdeen, Esther (April 1937). "The Sediments of Barataria Bay". Journal of Sedimentary Petrology. 7 (1). Retrieved 11 May 2014. (subscription required (help)).
6. "ISO 14688-1:2002 – Geotechnical investigation and testing – Identification and classification of soil – Part 1: Identification and description". International Organization for Standardization (ISO).
7. Folk, Robert L.; Ward, William C. (1957). "Brazos River bar: a study in the significance of grain-size parameters" (PDF). Journal of Sedimentary Petrology. 27 (1): 3–26. doi:10.1306/74d70646-2b21-11d7-8648000102c1865d. Retrieved 11 May 2014.
8. Southard, J B (1991). "Experimental Determination of Bed-Form Stability". Annual Review of Earth and Planetary Sciences. 19: 423–455. Bibcode:1991AREPS..19..423S. doi:10.1146/annurev.ea.19.050191.002231.
9. Perillo, Mauricio M. (2013). Flow, sediment transport and bedforms under combined flows (Ph.D.). University of Illinois at Urbana-Champaign.
10. Willmarth, W. W.; Lu, S. S. (1972). "Structure of the reynolds stress near the wall". Journal of Fluid Mechanics. 55 (1): 65–92. Bibcode:1972JFM....55...65W. doi:10.1017/S002211207200165X.
11. Lu, S. S.; Willmarth, W. W. (1973). "Measurements of the structure of the reynolds stress in a turbulent boundary layer". Journal of Fluid Mechanics. 60 (3): 481–511. Bibcode:1973JFM....60..481L. doi:10.1017/S0022112073000315.
12. Grass, A. J. (1983). "The influence of boundary layer turbulence on the mechanics of sediment transport". In Sumer, B. M.; Muller, A. Mechanics of Sediment Transport. A. A. Balkema. pp. 3–18.
13. Best, J. L. (1992). "On the entrainment of sediment and initiation of bed

defects: insights from recent developments within turbulent boundary layer research". Sedimentology. 39 (5): 797–811. Bibcode:1992Sedim..39..797B. doi:10.1111/j.1365-3091.1992.tb02154.x.

14. Venditti, J. G.; Church, M. A.; Bennett, S. J. (2005). "Bed form initiation from a flat sand bed". Journal of Geophysical Research. 110: F01009. Bibcode:2005JGRF..110.1009V. doi:10.1029/2004jf000149.
15. Venditti, J. G.; Church, M. A.; Bennett, S. J. (2006). "On interfacial instability as a cause of transverse subcritical bed forms". Water Resources Research. 42: W07423. Bibcode:2006WRR....42.7423V. doi:10.1029/2005wr004346.
16. Perillo, Mauricio M.; Best, James L.; Garcia, Marcelo H. (2014). "A new phase diagram for combined-flow bedforms.". Journal of Sedimentary Research. 84: 301–313. Bibcode:2014JSedR..84..301P. doi:10.2110/jsr.2014.25.
17. Prothero, D. R. and Schwab, F., 1996, Sedimentary Geology, pg. 45-49, ISBN 0-7167-2726-9
18. Klaus K.E. Neuendorf, James P. Mehl, Jr., Julia A. Jackson., eds. (2005). Glossary of geology. Alexandria: American Geological Institute. p. 382. ISBN 0-922152-76-4.
19. Celoria, Francis (1966). "Delta as a geographical concept in Greek literature". Isis. 57 (3): 385–388. doi:10.1086/350146.
20. Slingerland, R. and N. D. Smith (1998), "Necessary conditions for a meandering-river avulsion," Geology (Boulder), 26, 435–438.
21. Galloway, W.E., 1975, Process framework for describing the morphologic and stratigraphic evolution of deltaic depositional systems, in Brousard, M.L., ed., Deltas, Models for Exploration: Houston Geological Society, Houston, Texas, p. 87– 98.
22. Fagherazzi S., 2008, Self-organization of tidal deltas, Proceeding of the National Academy of Sciences, vol. 105 (48): 18692–18695,
23. Characteristics of deltas. (Available archived at – checked Dec 2008.)
24. Bernard Biju-Duval, J. Edwin Swezey. "Sedimentary Geology". Page 183. ISBN 2-7108-0802-1. Editions TECHNIP, 2002. Partial text on Google Books.
25. "Geological and Petrophysical Characterization of the Ferron Sandstone for 3-D Simulation of a Fluvial-deltaic Reservoir". By Thomas C. Chidsey, Thomas C. Chidsey, Jr (ed), Utah Geological Survey, 2002. ISBN 1-55791-668-3. Pages 2–17. Partial text on Google Books.
26. D.G.A Whitten, The Penguin Dictionary of Geology (1972)
27. Robert L. Bates, Julia A. Jackson, Dictionary of Geological Terms AGI (1984)
28. Hori, K. and Saito, Y. Morphology and Sediments of Large River Deltas. Tokyo, Japan: Tokyo Geographical Society, 2003
29. Mineral Information Institute. 2011. Sand and Gravel. http://www.mii.org/Minerals/photosandgr.html
30. Irwin III, R. et al. 2005. An intense terminal epoch of widespread fluvial

activity on early Mars: 2. Increased runoff and paleolake development. Journal of Geophysical Research: 10. E12S15

31. Coe, H.S.; Clevenger, G.H. (1916). "Methods for determining the capacities of slime-settling tanks". Transactions of the American Institute of Mining and Metallurgical Engineers. 55: 356.
32. "Siltation & Sedimentation". blackwarriorriver.org. Retrieved 2009-11-16.
33. "Siltation killed fish at Batang Rajang - Digest on Malaysian News". malaysiadigest.blogspot.com. Retrieved 2009-11-16.
34. U.D. Kulkarni; et al. "The International Journal of Climate Change: Impacts and Responses " Rate of Siltation in Wular Lake, (Jammu and Kashmir, India) with Special Emphasis on its Climate & Tectonics ". The International Journal of Climate Change: Impacts and Responses. Retrieved 2009-11-16.

CHAPTER 2

Sedimentary Structures

Chapter Contents

Sedimentary structures are visible features within sedimentary rocks that formed at the time of deposition and represent manifestations of the physical and biological processes that operated in depositional environments. The most commonly observed sedimentary structure is stratification. Stratification refers to the layering that occurs in sedimentary rocks as a result of different depositional episodes. Stratification is described by the thickness of the individual layers. Beds are layers of sediment that are thicker than 1 centimeter and are typically produced by coarser grained sediments such as sands and gravel. Laminations are layers thinner than 1 centimeter. These deposits are typically formed by fine grained sediments such as silts and clays and often represent seasonal deposition.

Other sedimentary structures provide information about specific processes operating within depositional environments. These structures are observed within individual beds or laminations and can be quite useful in determining environmental conditions at the time of deposition.

INTRODUCTION

Sedimentary structures are those structures formed during sediment deposition.

Sedimentary structures such as cross bedding, graded bedding and ripple marks are utilized in stratigraphic studies to indicate original position of strata in geologically complex terrains and understand the depositional environment of the sediment.

Ripple marks in a siltstone (Carmel Formation, Middle Jurassic, near Gunlock, Utah).

FLOW STRUCTURES

Megaripple/dune, formed in the upper flow regime, from Utah.

There are two kinds of flow structures: bidirectional (multiple directions, back-and-forth) and unidirectional. Flow regimes in single-direction (typically fluvial) flow, which at varying speeds and velocities produce different structures, are called bedforms.

In the *lower flow regime*, the natural progression is from a flat bed, to some sediment movement (saltation etc.), to ripples, to slightly larger dunes. Dunes have a vortex in the lee side of the dune. As the *upper flow regime* forms, the dunes become flattened out, and then produce antidunes. At higher still velocity, the antidunes are flattened and most sedimentation stops, as erosion takes over as the dominant process.

Bedforms vs. Flow

Typical unidirectional bedforms represent a specific flow velocity, assuming typical sediments (sands and silts) and water depths, and a chart such as below can be used for interpreting depositional environments, with **increasing** water velocity going down the chart.

Flow Regime	Bedform	Preservation Potential	Identification Tips
Lower			
	Lower plane bed	High	Flat laminae, almost lack of current
	Ripple marks	Relatively Low	Small, cm-scale undulations
	Sand waves	Medium to low	Rare, longer wavelength than ripples
	Dunes/Megaripples	Low	Large, meter-scale ripples
Upper			
	Upper plane bed	High	Flat laminae, +/- aligned grains (parting lineations)
	Antidunes	Low	Water in phase with bedform, low angle, subtle laminae
	Pool and chute	Very low	Mostly erosional features

RIPPLE MARKS

Wave ripple or symmetric ripple, from Permian rocks in Nomgon, Mongolia. Note "decapitation" of ripple crests due to change in current.

Ripple marks usually form in conditions with flowing water, in the lower part of the Lower Flow Regime. There are two types of ripple marks:

- Symmetrical ripple marks - Often found on beaches, they are created by a two way current, for example the waves on a beach (swash and backwash). This creates ripple marks with pointed crests and rounded troughs, which aren't inclined more to a certain direction. Three common sedimentary structures that are created by these processes are herringbone cross-stratification, flaser bedding, and interference ripples.
- Asymmetrical ripple marks - These are created by a one way current, for example in a river, or the wind in a desert. This creates ripple marks with still pointed crests and rounded troughs, but which are inclined more strongly in the direction of the current. For this reason, they can be used as palaeocurrent indicators.

ANTIDUNES

Antidunes are the sediment bedforms created by fast, shallow flows of water with a Froude number greater than 1. Antidunes form beneath standing waves of water that periodically steepen, migrate, and then break upstream. The antidune bedform is characterized by shallow foresets, which dip upstream at an angle of about ten degrees that can be up to five meters in length. They can be identified by their low angle foresets. For the most part, antidunes bedforms are destroyed during decreased flow, and therefore cross bedding formed by antidunes will not be preserved.

BIOLOGICAL STRUCTURES

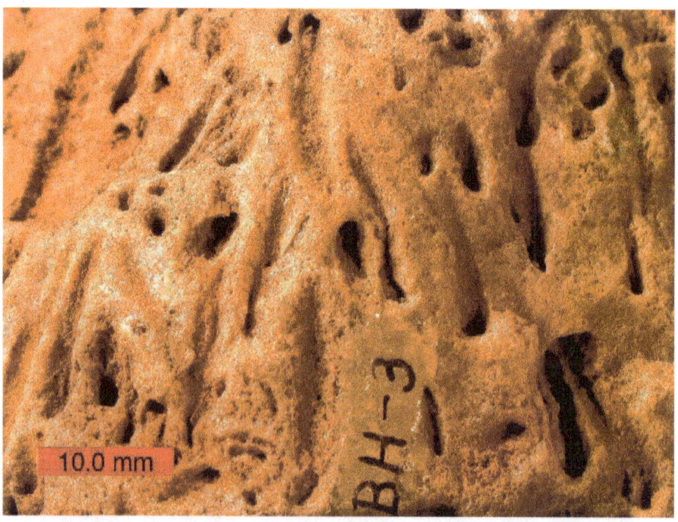

Skolithos trace fossil. Scale bar is 10 mm.

A number of biologically-created sedimentary structures exist, called trace fossils. Examples include burrows and various expressions of bioturbation. Ichnofacies are groups of trace fossils that together help give information on the depositional environment. In general, as deeper (into the sediment) burrows become more common, the shallower the water. As (intricate) surface traces become more common, the water becomes deeper.

Microbes may also interact with sediment to form Microbially Induced Sedimentary Structures.

SOFT SEDIMENT DEFORMATION STRUCTURES

Soft-sediment deformation structures develop at deposition or shortly after, during the first stages of the sediment's consolidation. This is because the sediments need to be "liquid-like" or unsolidified for the deformation to occur. These formations have also been put into a category called water-escape structures by Lowe (1975). The most common places for soft-sediment deformations to materialize are in deep water basins with turbidity currents, rivers, deltas, and shallow-marine areas with storm impacted conditions. This is because these environments have high deposition rates, which allows the sediments to pack loosely.

Soft sediment deformation in Dead Sea sediments, Israel. Possibly a seismite.

There are three common variants of SSD:

- load structures or load casts (also a type of sole marking) are blobs that form when a denser, wet sediment slumps down on and into a less dense sediment below.
- pseudonodules or ball-and-pillow structures, are pinched-off load structures; these may also be formed by earthquake energy and referred to as seismites.
- flame structures, "fingers" of mud that protrude into overlying sediments.

- clastic dikes are seams of sedimentary material that cut across sedimentary strata.

Cross-sectional view of deformed beds caused by soft-sediment deformation in the Booti Booti Sandstone (Mississippian), New South Wales. (Rygel, M.C.)

Aztec Sandstone (Lower Jurassic) in southern Nevada showing distorted eolian sand beds.

Large soft sediment deformation structures in turbidites, SE Spain. These are probably best described as flame structures, or perhaps ball-and-pillow structures. Backpack is around half a meter high. A small reverse fault runs through the outcrop in the center of the image.

Types of Soft-Sediment Deformation Structures

- Convolute bedding forms when complex folding and crumpling of beds or laminations occur. This type of deformation is found in fine or silty sands, and is usually confined to one rock layer. Convolute laminations are found in flood plain, delta, point-bar, and intertidal-flat deposits. They generally range in size from 3 to 25 cm, but there have been larger formations recorded as several meters thick.
- Flame structures consist of mud and are wavy or "flame" shaped. These flames usually extend into an overlying sandstone layer. This deformation is caused from sand being deposited onto mud, which is less dense. Load casts, technically a subset of sole markings, below, are the features which form alongside flame structures. Flames are thin fingers of mud injected upward into the overlying sands, while load casts are the pendulous knobs of sand that descend downwards into the mud between the flames.
- Slump structures are mainly found in sandy shales and mudstones, but may also be in limestones, sandstones, and evaporites. They are a result of the displacement and movement of unconsolidated sediments, and are found in areas with steep slopes and fast sedimentation rates. These structures often are faulted.
- Dish structures are thin, dish-shaped formations that normally occur in siltstones and sandstones. The size of each "dish" often ranges from 1 cm to 50 cm in size, and forms as a result of dewatering. Pillar structures often appear along with dish structures and also form by dewatering. They have a vertical orientation, which cuts across laminated or massive sands. These formations can range from a few millimeters in diameter to larger than a meter.

- Sole markings are found on the underside of sedimentary rocks that overlie shale beds, usually sandstones. They are used for determining the flow direction of old currents because of their directional features. Sole markings form from the erosion of a bed, which creates a groove that is later filled in by sediment.
- Seismites are sedimentary beds disturbed by seismic waves from earthquakes. They are commonly used to interpret the seismic history of an area. The term has also been applied to soft sediment deformation structures, including sand volcanos, sand blows, and certain clastic dikes.

BEDDING PLANE STRUCTURES

Flute cast from Book Cliffs area, Utah.

Mudcracks in rock at Roundtop Hill, Maryland.

Bedding Plane Structures are commonly used as paleocurrent indicators. They are formed when sediment has been deposited and then reworked and reshaped. They include:

- Sole markings form when an object gouges the surface of a sedimentary layer; this groove is later preserved as a cast when filled in by the layer above. They include:
    - Flute casts are scours dug into soft, fine sediment which typically get filled by an overlying bed. Measuring the long axis of the flute cast gives the direction of flow, with the scoop-shaped end pointing in the upcurrent direction and the tapered end pointing downcurrent (paleoflow direction). The convexity of the flute cast also points stratigraphically down.
    - Tool marks are a type of sole marking formed by grooves left in a bed by objects dragged along by a current. The average direction of these can be assumed to be the axis of flow direction.
- Mudcracks form when mud is dewatered, shrinks, and leaves a crack. This tells you that the mud was saturated with water and then exposed to air. Mudcracks curl upwards, so they can be used as geopetal structures. Syneresis cracks form in a similar way, with the exception that they are never exposed to air, instead being caused by changes in the salinity of the surrounding water.
- Raindrop impressions form on exposed sediment by raindrop impacts.
- Parting lineations are subtly aligned minerals that form in the lower part of the Upper Flow Regime within plane beds.

Within Bedding Structures

Cross-bedding and scour in a fine sandstone (Logan Formation, Mississippian, Jackson County, Ohio).

A teepee structure in modern halite deposits along the western shore of the Dead Sea, Israel.

These structures are within sedimentary bedding and can help with the interpretation of depositional environment and paleocurrent directions. They are formed when the sediment is deposited.

- Cross bedding - This can include ripples and dunes, or any cross stratification caused by currents. The "cross" refers to the angle between flat bedding and the inclined bedding of the cross bed, typically about 34 degrees. Paleocurrents are best found from cross beds that have 3D architecture exposed so you can measure the axis of the trough of the cross bed.
- Hummocky cross-stratification is made up of undulating sets of cross-laminae that are concave-up (swales) and convex-up (hummocks). These cross-beds gently cut into each other with curved erosional surfaces. They form in shallow-water, storm-dominated environments. Strong storm-wave action erodes the seabed into low hummocks and swales that lack a specific orientation.
- Imbrication is the stacking of larger clasts in the direction of flow.
- Normal graded bedding occurs when current velocity changes and grains are progressively dropped out of the current. The most common place to find this is in a turbidite deposit. This can also be inverted, called reversed graded bedding, and is common in debris flows.
- Bioturbation - Biological stirring of sediment (i.e. burrowing); typical of shallow water, finer-grained sediment.
- Tidal bundle - variation in bedding thickness in a tidal environment caused by alternation of spring and neap tides

SECONDARY SEDIMENTARY STRUCTURES

Secondary sedimentary structures form after primary deposition occurs or, in some cases, during the diagenesis of a sedimentary rock. Common secondary structures include any form of bioturbation, soft-sediment deformation, teepee structures, root-traces, and soil mottling. Liesegang rings, cone-in-cone structures, raindrop impressions, and vegetation-induced sedimentary structures would also be considered secondary structures.

REFERENCES

1. AILSA ALLABY and MICHAEL ALLABY. "sediment." A Dictionary of Earth Sciences. 1999. Encyclopedia.com. 8 Nov. 2010 <http://www.encyclopedia.com>.
2. Boggs, Sam jr, 2006 Principles of Sedimentology and Stratigraphy, Patrick Lynch, Principles of Sedimentology and Stratigraphy, Pearson Prentice Hall, Upper Saddle River, NJ. Ed 4 p 83-84
3. AILSA ALLABY and MICHAEL ALLABY. "antidune." A Dictionary of Earth Sciences. 1999. Encyclopedia.com. 8 Nov. 2010 <http://www.encyclopedia.com>. B
4. http://jsedres.geoscienceworld.org/cgi/content/abstract/35/4/922 C
5. Allen, John R. L. Sedimentary Structures, Their Character and Physical Basis. Vol. 2. Amsterdam: Elsevier Scientific Pub., 1982. Print
6. Boggs, S., 2006, Principles of Sedimentology and Stratigraphy (4th ed.), Pearson Prentice Hall, Upper Saddle River, NJ, p.94-97
7. Boggs, S., 2006, Principles of Sedimentology and Stratigraphy (4th ed.), Pearson Prentice Hall, Upper Saddle River, NJ, p.98-99
8. Seilacher, A., 1969, Fault-graded beds interpreted as seismites, Sedimentology, 13, p. 15-159
9. Seilacher, A., 1984, Sedimentary structures tentatively attributed to seismic events, Marine Geology, 55, p. 1-12
10. Bachmann, G.H., Aref, M.A.M., 2005, "A seismite in Triassic gypsum deposits (Grabfeld Formation, Ladinian), southwestern Germany", Sedimentary Geology, 180, p. 75–89
11. Ettensohn, F.R.; Rast, N.; Brett, C.E. (editors), Ancient Seismites, GSA Special Paper, 359

CHAPTER 3

Sedimentary Rocks

Chapter Contents

- Introduction
- Classification Based on Origin
- Deposition and Transformation
- Properties
- Sedimentary Environments
- Sedimentary Basins

Sedimentary rocks are formed by the deposition of mud, clay, sand and other small particles on the surface of the Earth. Sedimentation can occur in many different environments, including lakes, rivers, deserts and oceans. The sediment is deposited in layers that are compressed over time and turned into solid rock.

The most common types of sedimentary rocks are made from quartz or clay that is eroded and carried away by wind or water. These small particles are then deposited in a different part of the Earth. Over time, as these layers of deposition are buried and compressed, the heat and pressure fuse them together to form layers of solid rock, such as sandstone or shale.

Sedimentary rocks can also be formed from the remains of marine invertebrates, such as plankton and other small organisms. Their shells are made out of calcium carbonate. When they die, their remains sink slowly to the bottom of the ocean, forming thick layers that become limestone or chert.

Evaporating minerals from the ocean can also lead to the formation of sedimentary rocks. As shallow seas evaporate, layers of calcite, gypsum and halite settle out of solution. Over time, if these layers are left undisturbed, they can be turned into rocks known as evaporites.

INTRODUCTION

Sedimentary rocks are types of rock that are formed by the deposition and subsequent cementation of that material at the Earth's surface and within bodies of water. Sedimentation is the collective name for processes that cause mineral and/or organic particles (detritus) to settle in place. The particles that form a sedimentary rock by accumulating are called sediment. Before being deposited, the sediment was formed by weathering and erosion from the source area, and then transported to the place of deposition by water, wind, ice, mass movement or glaciers, which are called agents of denudation. Sedimentation may also occur as minerals precipitate from water solution or shells of aquatic creatures settle out of suspension.

The sedimentary rock cover of the continents of the Earth's crust is extensive (73% of the Earth's current land surface), but the total contribution of sedimentary rocks is estimated to be only 8% of the total volume of the crust. Sedimentary rocks are only a thin veneer over a crust consisting mainly of igneous and metamorphic rocks. Sedimentary rocks are deposited in layers as strata, forming a structure called bedding. The study of sedimentary rocks and rock strata provides information about the subsurface that is useful for civil engineering, for example in the construction of roads, houses, tunnels, canals or other structures. Sedimentary rocks are also important sources of natural resources like coal, fossil fuels, drinking water or ores.

The study of the sequence of sedimentary rock strata is the main source for an understanding of the Earth's history, including palaeogeography, paleoclimatology and the history of life. The scientific discipline that studies the properties and origin of sedimentary rocks is called sedimentology. Sedimentology is part of both geology and physical geography and overlaps partly with other disciplines in the Earth sciences, such as pedology, geomorphology, geochemistry and structural geology. Sedimentary rocks have also been found on Mars.

Middle Triassic marginal marine sequence of siltstones (reddish layers at the cliff base) and limestones (brown rocks above), Virgin Formation, southwestern Utah, USA

Sedimentary rocks on Mars, investigated by NASA's Curiosity Mars rover.

Steeply dipping sedimentary rock strata along the Chalous Road in northern Iran.

CLASSIFICATION BASED ON ORIGIN

Sedimentary rocks can be subdivided into four groups based on the processes responsible for their formation: clastic sedimentary rocks, biochemical (biogenic) sedimentary rocks, chemical sedimentary rocks, and a fourth category for "other" sedimentary rocks formed by impacts, volcanism, and other minor processes.

Clastic Sedimentary Rocks

Clastic sedimentary rocks are composed of other rock fragments that were cemented by silicate minerals. Clastic rocks are composed largely of quartz, feldspar, rock (lithic) fragments, clay minerals, and mica; any type of mineral may be present, but they in general represent the minerals that exist locally.

Clastic sedimentary rocks, are subdivided according to the dominant particle size. Most geologists use the Udden-Wentworth grain size scale and divide unconsolidated sediment into three fractions: gravel (>2 mm diameter), sand (1/16 to 2 mm diameter), and mud (clay is <1/256 mm and silt is between 1/16 and 1/256 mm). The classification of clastic sedimentary rocks parallels this scheme; conglomerates and breccias are made mostly of gravel, sandstones are made mostly of sand, and mudrocks are made mostly of the finest material. This tripartite subdivision is mirrored by the broad categories of rudites, arenites, and lutites, respectively, in older literature.

The subdivision of these three broad categories is based on differences in clast shape:conglomerates and breccias), composition (sandstones), grain size and/or texture (mudrocks).

Claystone deposited in Glacial Lake Missoula, Montana, United States. Note the very fine and flat bedding, common for distal lacustrine deposition.

Diagenesis of Siliciclastic Sedimentary Rocks

Siliciclastic rocks initially form as loosely packed sediment deposits including gravels, sands, and muds. The process of turning loose sediment into hard is called lithification. During the process of lithification, sediments undergo physical, chemical and mineralogical changes before becoming rock. The primary physical process in lithification is compaction. As sediment transport and deposition continues, new sediments are deposited over top of previously deposited beds burying them. Burial continues and the weight of overlying sediments cause an increase in temperature and pressure. This increase in temperature and pressure causes loose grained sediments become tightly packed reducing porosity, essentially squeezing water out of the sediment. Porosity is further reduced by the precipitation of minerals into the remaining pore spaces. The final stage in the process is diagenesis and will be discussed in detail below.

Cementation

Cementation is the diagenetic process by which coarse clastic sediments become lithified or consolidated into hard, compact rocks, usually through the deposition or precipitation of minerals in the spaces between the individual grains of sediment. Cementation can occur simultaneously with deposition or at another time. Furthermore, once a sediment is deposited, it becomes subject to cementation through the various stages of diagenesis discussed below.

Shallow Burial (Eogenesis)

Eogenesis refers to the early stages of diagenesis. This can take place at very shallow depths, ranging from a few meters to tens of meters below the surface. The changes that occur during this diagenetic phase mainly relate to the reworking of the sediments. Compaction and grain repacking, bioturbation, as well as mineralogical changes all occur at varying degrees. Due to the shallow depths, sediments undergo only minor compaction and grain rearrangement during this stage. Organisms rework sediment near the depositional interface by burrowing, crawling, and in some cases sediment ingestion. This process can destroy sedimentary structures that were present upon deposition of the sediment. Structures such as lamination will give way to new structures associated the activity of organisms. Despite being close to the surface, eogenesis does provide conditions for important mineralogical changes to occur. This mainly involves the precipitation of new minerals.

Mineralogical Changes During Eogenesis

Mineralogical changes that occur during eogenesis as dependent on the environment in which that sediment has been deposited. For example, the formation of pyrite is characteristic of reducing conditions in marine environments. Pyrite can form as cement, or replace organic materials, such as wood fragments. Other important reactions include the formation of chlorite, glauconite, illite and iron oxide (if oxygenated pore water is present). The precipitation of potassium feldspar, quartz overgrowths, and carbonate cements also occurs under marine conditions. In non marine environments oxidizing conditions are almost always prevalent, meaning iron oxides are commonly produced along with kaolin group clay minerals. The precipitation of quartz and calcite cements may also occur in non marine conditions.

Deep Burial (Mesogenesis)

Compaction

As sediments are buried deeper, load pressures become greater resulting in tight grain packing and bed thinning. This causes increased pressure between grains thus increasing the solubility of grains. As a result, the partial dissolution of silicate grains occurs. This is called pressure solutions. Chemically speaking, increases in temperature can also cause chemical reaction rates to increase. This increases the solubility of most common minerals (aside from evaporites). Furthermore, beds thin and porosity decreases allowing cementation to occur by the precipitation of silica or carbonate cements into remaining pore space.

In this process minerals crystallize from watery solutions that percolate through the pores between grain of sediment. The cement that is produced may or may not have the same chemical composition as the sediment. In sandstones, framework grains are often cemented by silica or carbonate. The extent of cementation is dependent on the composition of the sediment. For example, in lithic sandstones, cementation is less extensive because pore space between framework grains is filled with a muddy matrix that leaves little space for precipitation to occur. This is often the case for mudrocks as well. As a result of compaction, the clayey sediments comprising mudrocks are relatively impermeable.

Dissolution

Dissolution of framework silicate grains and previously formed carbonate cement may occur during deep burial. Conditions that encourage this are essentially opposite of those required for cementation. Rock fragments and silicate minerals of low stability, such as plagioclase feldspar, pyroxenes, and amphiboles, may dissolve as a result of increasing burial temperatures and the presence of organic acids in pore waters. The dissolution of frame work grains and cements increases porosity particularly in sandstones.

Mineral Replacement

This refers to the process whereby one mineral is dissolved and a new mineral fills the space via precipitation. Replacement can be partial or complete. Complete replacement destroys the identity of the original minerals or rock fragments giving a biased view of the original mineralogy of the rock/ Porosity can also be affected by this process. For example, clay minerals tend to fill up pore space and thereby reducing porosity.

Telogenesis

In the process of burial, it is possible that siliciclastic deposits may subsequently be uplifted as a result of a mountain building event or erosion. When uplift occurs, it exposes buried deposits to a radically new environment. Because the process brings material to or closer to the surface, sediments that undergo uplift are subjected to lower temperatures and pressures as well as slightly acidic rain water. Under these conditions, framework grains and cement are again subjected to dissolution and in turn increasing porosity. On the other hand, telogenesis can also change framework grains to clays, thus reducing porosity. These changes are dependent on the specific conditions that the rock is exposed as well as the composition of the rock and pore waters. Specific pore waters, can cause the further precipitation of carbonate or silica cements. This process can also encourage the process of oxidation on a variety of iron bearing minerals.

Conglomerates and Breccias

Conglomerates are dominantly composed of rounded gravel, while breccias are composed of dominantly angular gravel.

Conglomerate is a coarse-grained clastic sedimentary rock that is composed of a substantial fraction of rounded to subangular gravel-size clasts, e.g., granules, pebbles, cobbles, and boulders, larger than 2 mm (0.079 in) in diameter. Conglomerates form by the consolidation and lithification of gravel. Conglomerates typically contain finer grained sediment, e.g., either sand, silt, clay or combination of them, called *matrix* by geologists, filling their interstices and are often cemented by calcium carbonate, iron oxide, silica, or hardened clay.

The size and composition of the gravel-size fraction of a conglomerate may or may not vary in composition, sorting, and size. In some conglomerates, the gravel-size class consist almost entirely of what were clay clasts at the time of deposition. Conglomerates can be found in sedimentary rock sequences of all ages but probably make up less than 1 percent by weight of all sedimentary rocks. In terms of origin and depositional mechanisms, they are closely related to sandstones and exhibit many of the same types of sedimentary structures, e.g., tabular and trough cross-bedding and graded bedding.

Boulder of conglomerate with cobble-sized clasts. Rock hammer for scale.

Carmelo Formation (Conglomerate) at Point Lobos.

Classification

Conglomerates may be named and classified by the:

- Amount and type of matrix present
- Composition of gravel-size clasts they contain
- Size range of gravel-size clasts present

The classification method depends on the type and detail of research being conducted.

A sedimentary rock composed largely of gravel is first named according to the roundness of the gravel. If the gravel clasts that comprise it is largely well-rounded to subrounded, it is a *conglomerate.* If the gravel clasts that comprise it are largely angular, it is a breccia. Such breccias can be called sedimentary breccias to differentiate them from other types of breccia, e.g. volcanic and fault breccias. Sedimentary rocks that contain a mixture of rounded and angular gravel clasts are sometimes called breccio-conglomerate.

Texture

Conglomerates are rarely composed entirely of gravel-size clasts. Typically, the space between the gravel-size clasts is filled by a mixture composed of varying amounts of silt, sand, and clay, known as *matrix*. If the individual gravel clasts in a conglomerate are separated from each other by an abundance of matrix such that they are not in contact with each other and *float* within the matrix, it is called a paraconglomerate. Paraconglomerates are also often unstratified and can contain more matrix than gravel clasts. If the gravel clasts of a conglomerate are in contact with each other, it is called a orthoconglomerate. Unlike paraconglomerates, orthoconglomerates are typically cross-bedded and often well-cemented and lithified by either calcite, hematite, quartz, or clay.

The differences between paraconglomerates and orthoconglomerates reflect differences in how they are deposited. Paraconglomerates are commonly either glacial tills or debris flow deposits. Orthoconglomerates are tyipically associated with aqueous currents of some sort.

A conglomerate at the base of the Cambrian in the Black Hills, South Dakota.

Section of polymict conglomerate from offshore rock core, Alaska, approximate depth 10,000 ft.

Conglomerate and sandstone alternating layers - Nerriga, New South Wales, Australia. Approx 5m vertical wall of road cut (vertical marks by construction machine.)

Clast Composition

Conglomerates are also classified according to the composition of their clasts. A conglomerate or any clastic sedimentary rock that consists of a single rock or mineral is known as either a monomict, monomictic, oligomict, or oligomictic conglomerate. If the conglomerate consists of two or more different types of rocks, minerals, or combination of both, it is known as either a polymict or polymictic conglomerate. If a polymictic conglomerate contains an assortment of the clasts of metastable and unstable rocks and minerals, it called either a petromict or petromictic conglomerate.

In addition, conglomerates are classified by source as indicated by the lithology of the gravel-size clasts If these clasts consist of rocks and minerals that are significantly different in lithology from the enclosing matrix and, thus, older and derived from outside the basin of deposition, the conglomerate is known as an extraformational conglomerate. If these clasts consist of rocks and minerals that are identical to or consistent with the lithology of the enclosing matrix and, thus, penecontemporaneous and derived from within the basin of deposition, the conglomerate is known as an intraformational conglomerate.

Two recognized types of type of intraformational conglomerates are shale-pebble and flat-pebble conglomerates. A shale-pebble conglomerate is a conglomerate that is composed largely of clasts of rounded mud chips and pebbles held together by clay minerals and created by erosion within environments such as within a river channel or along a lake margin. Flat-pebble conglomerates (edgewise conglomerates) are conglomerates that consist of relatively flat clasts of lime mud created by either storms or tsunami eroding a shallow sea bottom or tidal currents eroding tidal flats along a shoreline.

Clast Size

Finally, conglomerates are often differentiated and named according to the dominant clast size comprising them. In this classification, a conglomerate composed largely of granule-size clasts would be called a granule conglomerate; a conglomerate composed largely of pebble-size clasts would be called a pebble conglomerate; and a conglomerate composed largely of cobble-size clasts would be called a cobble conglomerate.

Sedimentary Environments

Conglomerates are deposited in a variety of sedimentary environments.

Deepwater Marine

In turbidites, the basal part of a bed is typically coarse-grained and sometimes conglomeratic. In this setting, conglomerates are normally very well sorted, well-rounded and often with a strong A-axis type imbrication of the clasts.

Shallow Marine

Conglomerates are normally present at the base of sequences laid down during marine transgressions above an unconformity, and are known as *basal conglomerates*. They represent the position of the shoreline at a particular time and are diachronous.

Fluvial

Conglomerates deposited in fluvial environments are typically well rounded and well sorted. Clasts of this size are carried as bedload and only at times of high flow-rate. The maximum clast size decreases as the clasts are transported further due to attrition, so conglomerates are more characteristic of immature river systems. In the sediments deposited by mature rivers, conglomerates are generally confined to the basal part of a channel fill where they are known as *pebble lags*. Conglomerates deposited in a fluvial environment often have an AB-plane type imbrication.

Alluvial

Fanglomerate.

Alluvial deposits form in areas of high relief and are typically coarse-grained. At mountain fronts individual alluvial fans merge to form braidplains and these two environments are associated with the thickest deposits of conglomerates. The bulk of conglomerates deposited in this setting are clast-supported with a strong AB-plane imbrication. Matrix-supported conglomerates, as a result of debris-flow deposition, are quite commonly associated with many alluvial fans. When such conglomerates accumulate within an alluvial fan, in rapidly eroding (e.g., desert) environments, the resulting rock unit is often called a *fanglomerate.*

Glacial

Glaciers carry a lot of coarse-grained material and many glacial deposits are conglomeratic. Tillites, the sediments deposited directly by a glacier, are typically poorly sorted, matrix-supported conglomerates. The matrix is generally fine-grained, consisting of finely milled rock fragments. Waterlaid deposits associated with glaciers are often conglomeratic, forming structures such as eskers.

Examples

An example of conglomerate can be seen at Montserrat, near Barcelona. Here, erosion has created vertical channels that give the characteristic jagged shapes the mountain is named for (Montserrat literally means "jagged mountain"). The rock is strong enough to use as a building material, as in the Santa Maria de Montserrat Abbey.

Another example, the Crestone Conglomerate, occurs in and near the town of Crestone, at the foot of the Sangre de Cristo Range in Colorado's San Luis Valley. The Crestone Conglomerate consists of poorly sorted fanglomerates that accumulated in prehistoric alluvial fans and related fluvial systems. Some of these rocks have hues of red and green.

Conglomerate cliffs are found on the east coast of Scotland from Arbroath northwards along the coastlines of the former counties of Angus and Kincardineshire. Dunnottar Castle sits on a rugged promontory of conglomerate jutting into the North Sea just south of the town of Stonehaven.

Conglomerate may also be seen in the domed hills of Kata Tjuta, in Australia's Northern Territory.

In the nineteenth century a thick layer of Pottsville conglomerate was recognized to underlie anthracite coal measures in Pennsylvania.

Examples on Mars

On Mars, slabs of conglomerate have been found at an outcrop named "Hottah", and have been interpreted by scientists as having formed in an ancient streambed. The gravels, which were discovered by NASA's Mars rover Curiosity, range from the size of sand particles to the size of golf balls. Analysis has shown that the pebbles were deposited by a stream that flowed at walking pace and was ankle- to hip-deep.

Metaconglomerate

Metamorphic alteration transforms conglomerate into metaconglomerate.

Sandstones

Sandstone classification schemes vary widely, but most geologists have adopted the Dott scheme, which uses the relative abundance of quartz, feldspar, and lithic framework grains and the abundance of a muddy matrix between the larger grains.

Composition of framework grains

The relative abundance of sand-sized framework grains determines the first word in a sandstone name. Naming depends on the dominance of the three most abundant components quartz, feldspar, or the lithic fragments that originated from other rocks. All other minerals are considered accessories and not used in the naming of the rock, regardless of abundance.

- Quartz sandstones have >90% quartz grains
- Feldspathic sandstones have <90% quartz grains and more feldspar grains than lithic grains
- Lithic sandstones have <90% quartz grains and more lithic grains than feldspar grains

Abundance of muddy matrix material between sand grains

When sand-sized particles are deposited, the space between the grains either remains open or is filled with mud (silt and/or clay sized particle).

- "Clean" sandstones with open pore space (that may later be filled with matrix material) are called arenites.
- Muddy sandstones with abundant (>10%) muddy matrix are called wackes.

Sedimentary rock with sandstone in Malta.

Six sandstone names are possible using the descriptors for grain composition (quartz-, feldspathic-, and lithic-) and the amount of matrix (wacke or arenite). For example, a quartz arenite would be composed of mostly (>90%) quartz grains and have little or no clayey matrix between the grains, a lithic wacke would have abundant lithic grains and abundant muddy matrix, etc.

Although the Dott classification scheme is widely used by sedimentologists, common names like greywacke, arkose, and quartz sandstone are still widely used by non-specialists and in popular literature.

Components

Framework Grains

The Paradise Quarry in Sydney, Australia.

Grus sand and granitoid it derived from.

Framework grains are sand-sized (0.0625-to-2-millimetre (0.00246 to 0.07874 in) diameter) detrital fragments that make up the bulk of a sandstone. These grains can be classified into several different categories based on their mineral composition:

- Quartz framework grains are the dominant minerals in most clastic sedimentary rocks; this is because they have exceptional physical properties, such as hardness and chemical stability. These physical properties allow the quartz grains to survive multiple recycling events, while also allowing the grains to display some degree of rounding. Quartz grains evolve from plutonic rock, which are felsic in origin and also from older sandstones that have been recycled.
- Feldspathic framework grains are commonly the second most abundant mineral in sandstones. Feldspar can be divided into two smaller subdivisions: alkali feldspars and plagioclase feldspars. The different types of feldspar can be distinguished under a petrographic microscope. Below is a description of the different types of feldspar.
- Alkali feldspar is a group of minerals in which the chemical composition of the mineral can range from $KAlSi_3O_8$ to $NaAlSi_3O_8$, this represents a complete solid solution.
- Plagioclase feldspar is a complex group of solid solution minerals that range in composition from $NaAlSi_3O_8$ to $CaAl_2Si_2O_8$.
- Lithic framework grains are pieces of ancient source rock that have yet to weather away to individual mineral grains, called lithic fragments or clasts. Lithic fragments can be any fine-grained or coarse-grained igneous, metamorphic, or sedimentary rock, although the most common lithic fragments found in sedimentary rocks are clasts of volcanic rocks.
- Accessory minerals are all other mineral grains in a sandstone; commonly these minerals make up just a small percentage of the grains in a sandstone. Common accessory minerals include micas (muscovite and biotite), olivine, pyroxene, and corundum. Many of these accessory grains are more dense than the silicates that make up the bulk of the rock. These heavy minerals are commonly resistant to weathering and can be used as an indicator of sandstone maturity through the ZTR index. Common heavy minerals include zircon, tourmaline, rutile (hence *ZTR*), garnet, magnetite, or other dense, resistant minerals derived from the source rock.

Matrix

Matrix is very fine material, which is present within interstitial pore space between the framework grains. The interstitial pore space can be classified into two different varieties. One is to call the sandstone an arenite, and the other is to call it a wacke. Below is a definition of the differences between the two matrices.

- Arenites are texturally *clean* sandstones that are free of or have very little matrix.
- Wackes are texturally *dirty* sandstones that have a significant amount of matrix.

Cement

Cement is what binds the siliciclastic framework grains together. Cement is a secondary mineral that forms after deposition and during burial of the sandstone. These cementing materials may be either silicate minerals or non-silicate minerals, such as calcite.

- Silica cement can consist of either quartz or opal minerals. Quartz is the most common silicate mineral that acts as cement. In sandstone where there is silica cement present the quartz grains are attached to cement, this creates a rim around the quartz grain called overgrowth. The overgrowth retains the same crystallographic continuity of quartz framework grain that is being cemented. Opal cement is found in sandstones that are rich in volcanogenic materials, and very rarely is in other sandstones.
- Calcite cement is the most common carbonate cement. Calcite cement is an assortment of smaller calcite crystals. The cement adheres itself to the framework grains, this adhesion is what causes the framework grains to be adhered together.
- Other minerals that act as cements include: hematite, limonite, feldspars, anhydrite, gypsum, barite, clay minerals, and zeolite minerals.

Pore Space

Pore space includes the open spaces within a rock or a soil. The pore space in a rock has a direct relationship to the porosity and permeability of the rock. The porosity and permeability are directly influenced by the way the sand grains are packed together.

- Porosity is the percentage of bulk volume that is inhabited by interstices within a given rock. Porosity is directly influenced by the packing of even-sized spherical grains, rearranged from loosely packed to tightest packed in sandstones.
- Permeability is the rate in which water or other fluids flow through the rock. For groundwater work permeability may be measured in gallons per day through a one square foot cross section under a unit hydraulic gradient.

Types of Sandstone

All sandstones are composed of the same general minerals. These minerals make up the framework components of the sandstones. Such components are quartz, feldspars, and lithic fragments. Matrix may also be present in the interstitial spaces between the framework grains. Below is a list of several major groups of sandstones. These groups are divided based on mineralogy and texture. Even though sandstones have very simple compositions which are based on framework grains, geologists have not been able to agree on a specific, right way, to classify sandstones. Sandstone classifications are typically done by point-counting a thin section using a method like the Gazzi-Dickinson Method. The composition of a sandstone can have important information regarding the genesis of the sediment when used with a triangular *Q*uartz, *F*eldspar, *L*ithic fragment (QFL diagrams). Many geologists, however, do not agree on how to separate the triangle parts into the single components so that the framework grains can be plotted. Therefore, there have been many published ways to classify sandstones, all of which are similar in their general format.

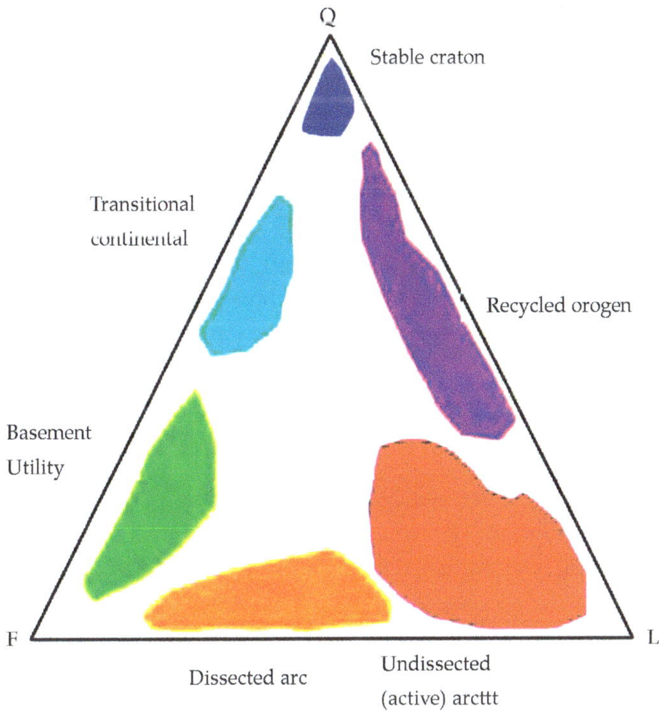

Schematic QFL diagram showing tectonic provinces.

Visual aids are diagrams that allow geologists to interpret different characteristics about a sandstone. The following QFL chart and the sandstone provenance model correspond with each other therefore, when the QFL chart is plotted those points can then be plotted on the sandstone provenance model. The stage of textural maturity chart illustrates the different stages that a sandstone goes through.

- A QFL chart is a representation of the framework grains and matrix that is present in a sandstone. This chart is similar to those used in igneous petrology. When plotted correctly, this model of analysis creates for a meaningful quantitative classification of sandstones.
- A sandstone provenance chart allows geologists to visually interpret the different types of places sandstones can originate from.
- A stage of textural maturity is a chart that shows the different stages of sandstones. This chart shows the difference between immature, submature, mature, and supermature sandstones. As the sandstone becomes more mature grains become more rounded, and there is less clay that makes up the matrix of the rock.

Dott Classification Scheme

Dott's (1964) sandstone classification scheme is one of many classification scheme used by geologists for classifying sandstones. Dott's scheme is a modification of Gilbert's classification of silicate sandstones, and it incorporates R.L. Folk's dual textural and compositional maturity concepts into one classification system. The philosophy behind combining Gilbert's classification scheme and R. L. Folk's classification scheme is that it is better able to "portray the continuous nature of textural variation from mudstone to arenite and from stable to unstable grain composition". Dott's classification scheme is based on the mineralogy of framework grains, and on the type of matrix present in between the framework grains.

In this specific classification scheme, Dott has set the boundary between arenite and wackes at 15% matrix. In addition to setting a boundary for what the matrix is, Dott also breaks up the different types framework grains that can be present in a sandstone into three major categories: quartz, feldspar, and lithic grains.

- Arenites are types of sandstone that have less than 15% clay matrix in between the framework grains.
    - Quartz arenites are sandstones that contain more than 90% of siliceous grains. Grains can include quartz or chert rock fragments. Quartz arenites are texturally mature to supermature sandstones. These pure quartz sands result from extensive weathering that occurred before and during transport. This weathering removed everything but quartz grains, the most stable mineral. They are commonly affiliated with rocks that are deposited in a stable cratonic environment, such as aeolian beaches or shelf environments. Quartz arenites emanate from multiple recycling of quarts grains, generally as sedimentary source rocks and less regularly as first-cycle deposits derived from primary igneous or metamorphic rocks.
    - Feldspathic arenites are sandstones that contain less than 90% quartz, and more feldspar than unstable lithic fragments, and minor accessory minerals. Feldspathic sandstones are commonly immature or sub-mature. These sandstones occur in association with cratonic or stable shelf settings. Feldspathic sandstones are derived from granitic-type, primary crystalline, rocks. If the sandstone is dominantly plagioclase, then it is igneous in origin.
    - Lithic arenites are characterised by generally high content of unstable lithic fragments. Examples include volcanic and metamorphic clasts, though stable clasts such as chert are common in lithic arenites. This type of rock contains less than 90% quartz grains and more unstable rock fragments than feldspars. They are commonly immature to submature texturally. They are associated with fluvial conglomerates and other fluvial deposits, or in deeper water marine conglomerates. They are formed under conditions that produce large volumes of unstable material, derived from fine-grained rocks, mostly shales, volcanic rocks, and metamorphic rock.

- Wackes are sandstones that contain more than 15% clay matrix in between framework grains.
    - Quartz wackes are uncommon because quartz arenites are texturally mature to supermature.
    - Felspathic wackes are feldspathic sandstone that contain a matrix that is greater than 15%.
    - Lithic wacke is a sandstone in which the matrix greater than 15%.
- Arkose sandstones are more than 25 percent feldspar. The grains tend to be poorly rounded and less well sorted than those of pure quartz sandstones. These feldspar-rich sandstones come from rapidly eroding granitic and metamorphic terrains where chemical weathering is subordinate to physical weathering.
- Greywacke sandstones are a heterogeneous mixture of lithic fragments and angular grains of quartz and feldspar or grains surrounded by a fine-grained clay matrix. Much of this matrix is formed by relatively soft fragments, such as shale and some volcanic rocks, that are chemically altered and physically compacted after deep burial of the sandstone formation.

Mudrocks

Mudrocks are a class of fine grained siliciclastic sedimentary rocks. The varying types of mudrocks include: *siltstone*, *claystone*, *mudstone*, *slate*, and *shale*. Most of the particles of which the stone is composed are less than 0.0625 mm (1/16th mm or 0.0025 inches) and are too small to study readily in the field. At first sight the rock types look quite similar; however, there are important differences in composition and nomenclature. There has been a great deal of disagreement involving the classification of mudrocks. There are a few important hurdles to classification, including:

- Mudrocks are the least understood, and one of the most understudied sedimentary rocks to date
- It is difficult to study mudrock constituents, due to their diminutive size and susceptibility to weathering on outcrops
- And most importantly, there is more than one classification scheme accepted by scientists

Mudrocks make up fifty percent of the sedimentary rocks in the geologic record, and are easily the most widespread deposits on Earth. Fine sediment is the most abundant product of erosion, and these sediments contribute to the overall omnipresence of mudrocks. With increased pressure over time the platey clay minerals may become aligned, with the appearance of parallel layering (fissility). This finely bedded material that splits readily into thin layers is called *shale*, as distinct from *mudstone*. The lack of fissility or layering in mudstone may be due either to the original texture or to the disruption of layering by burrowing organisms in the sediment prior to lithification.

From the beginning of civilization, when pottery and mudbricks were made by hand, to now, mudrocks have been important. The first book on mudrocks, *Geologie des Argils* by Millot, was not published until 1964; however, scientists, engineers, and oil producers have understood the significance of mudrocks since the discovery of the Burgess Shale and the relatedness of mudrocks and oil. Literature on this omnipresent rock-type has been increasing in recent years, and technology continues to allow for better analysis.

Red mudrock in the Ragged Reef Formation (Pennsylvanian), Cumberland Basin, Nova Scotia.

Nomenclature

Mudrocks, by definition, consist of at least fifty percent mud-sized particles. Specifically, mud is composed of silt-sized particles that are between 1/16 – 1/256 ($(1/16)^2$) of a millimeter in diameter, and clay-sized particles which are less than 1/256 millimeter.

Mudrocks contain mostly clay minerals, and quartz and feldspars. They can also contain the following particles at less than 63 micrometres: calcite, dolomite, siderite, pyrite, marcasite, heavy minerals, and even organic carbon.

There are various synonyms for fine-grained siliciclastic rocks containing fifty percent or more of its constituents less than 1/256 of a millimeter. Mudstones, shales, lutites, and argillites are common qualifiers, or umbrella-terms; however, the term mudrock has increasingly become the terminology of choice by sedimentary geologists and authors.

The term "mudrock" allows for further subdivisions of siltstone, claystone, mudstone, and shale. For example, a siltstone would be made of more than 50-percent grains that equate to 1/16 - 1/256 of a millimeter. "Shale" denotes fissility, which implies an ability to part easily or break parallel to stratification. Siltstone, mudstone, and claystone implies lithified, or hardened, detritus without fissility.

Overall, "mudrocks" may be the most useful qualifying term, because it allows for rocks to be divided by its greatest portion of contributing grains and their respective grain size, whether silt, clay, or mud.

Type	Min grain	Max grain
Claystone	0 μm	4 μm
Mudstone	0 μm	64 μm
Siltstone	4 μm	64 μm
Shale	0 μm	64 μm
Slate	na	na

Claystone

A claystone is lithified, and non-fissile mudrock. In order for a rock to be considered a claystone, it must consist of up to fifty percent clay, which measures less than 1/256 of a millimeter in particle size. Clay minerals are integral to mudrocks, and represent the first or second most abundant constituent by volume. There are 35 recognized clay mineral species on Earth. They make muds cohesive and plastic, or able to flow. Clay is by far the smallest particles recognized in mudrocks. Most materials in nature are clay minerals, but quartz, feldspar, iron oxides, and carbonates can weather to sizes of a typical clay mineral.

Claystone in Slovakia.

For a size comparison, a clay-sized particle is 1/1000 the size of a sand grain. This means a clay particle will travel 1000 times further at constant water velocity, thus requiring quieter conditions for settlement.

The formation of clay is well understood, and can come from soil, volcanic ash, and glaciation. Ancient mudrocks are another source, because they weather and disintegrate easily. Feldspar, amphiboles, pyroxenes, and volcanic glass are the principle donors of clay minerals.

Mudstone

A mudstone is a siliciclastic sedimentary rock that contains a mixture of silt- and clay-sized particles (at least 1/3 of each).

The terminology of "mudstone" is not to be confused with the Dunham classification scheme for limestones. In Dunham's classification, a mudstone is any limestone containing less than ten percent carbonate grains. Note, a siliciclastic mudstone does not deal with carbonate grains. Friedman, Sanders, and Kopaska-Merkel (1992) suggest the use of "lime mudstone" to avoid confusion with siliciclastic rocks.

Siltstone

A siltstone is a lithified, non-fissile mudrock. In order for a rock to be named a siltstone, it must contain over fifty percent silt-sized material. Silt is any particle smaller than sand, 1/16 of a millimeter, and larger than clay, 1/256 of millimeter. Silt is believed to be the product of physical weathering, which can involve freezing and thawing, thermal expansion, and release of pressure. Physical weathering does not involve any chemical changes in the rock, and it may be best summarised as the physical breaking apart of a rock.

Siltstone at UAT, Estonia.

One of the highest proportions of silt found on Earth is in the Himalayas, where phyllites are exposed to rainfall of up to five to ten meters (16 to 33 feet) a year. Quartz and feldspar are the biggest contributors to the silt realm, and silt tends to be non-cohesive, non-plastic, but can liquefy easily.

There is a simple test that can be done in the field to determine whether a rock is a siltstone or not, and that is to put the rock to one's teeth. If the rock feels "gritty" against one's teeth, then it is a siltstone.

Shale

Shale is a fine grained, hard, laminated mudrock, consisting of clay minerals, and quartz and feldspar silt. Shale is lithified and fissile. It must have at least 50-percent of its particles measure less than 0.062 mm. This term is confined to argillaceous, or clay-bearing, rock.

Marcellus Shale, New York.

Black Shale with pyrite.

There are many varieties of shale, including calcareous and organic-rich; however, black shale, or organic-rich shale, deserves further evaluation. In order for a shale to be a black shale, it must contain more than one percent organic carbon. A good source rock for hydrocarbons can contain up to twenty percent organic carbon. Generally, black shale receives its influx of carbon from algae, which decays and forms an ooze known as sapropel. When this ooze is cooked at desired pressure, three to six kilometers (1.8 - 3.7 miles) depth, and temperature, 90–120 °C (194–248 °F), it will form kerogen. Kerogen can be heated, and yield up to 10–150 US gallons (0.038–0.568 m^3) of product per ton of rock.

Slate

Slate is a hard mudstone that has undergone metamorphosis, and has well-developed cleavage. It has gone through metamorphism at temperatures between 200–250 °C (392–482 °F), or extreme deformation. Since slate is formed in the lower realm of metamorphism, based on pressure and temperature, slate retains its stratification and can be defined as a hard, fine-grained rock.

Slate Roof.

Slate is often used for roofing, flooring, or old-fashioned stone walls. It has an attractive appearance, and its ideal cleavage and smooth texture are desirable.

Creation of Mud and Mudrocks

Most mudrocks form in oceans or lakes, because these environments provide the quiet waters necessary for deposition. Although mudrocks can be found in every depositional environment on Earth, the majority are found in lakes and oceans.

Mud Transport and Supply

Heavy rainfall provides the kinetic motion necessary for mud, clay, and silt transport. Southeast Asia, including Bangladesh and India, receives high amounts of rain from monsoons, which then washes sediment from the Himalayas and surrounding areas to the Indian Ocean.

Warm, wet climates are best for weathering rocks, and there is more mud on ocean shelves off tropical coasts than on temperate or polar shelves. The Amazon system, for example, has the third largest sediment load on Earth, with rainfall providing clay, silt, and mud from the Andes in Peru, Ecuador, and Bolivia.

Rivers, waves, and longshore currents segregate mud, silt, and clay from sand and gravel due to fall velocity. Longer rivers, with low gradients and large watersheds, have the best carrying capacity for mud. The Mississippi River, a good example of long, low gradient river with a large amount of water, will carry mud from its northernmost sections, and deposit the material in its mud-dominated delta.

Mudrock Depositional Environments

Below is a listing of various environments that act as sources, modes of transportation to the oceans, and environments of deposition for mudrocks.

Alluvial Environments

Alluvial valleys, such as the Ganges in India, the Yellow in China, and Lower Mississippi in the United States are good examples of alluvial valleys. These systems have a continuous source of water, and can contribute mud through overbank sedimentation, when mud and silt is deposited overbank during flooding, and oxbow sedimentation where an abandoned stream is filled by mud.

In order for an alluvial valley to exist there must be a highly elevated zone, usually uplifted by active tectonic movement, and a lower zone, which acts as a conduit for water and sediment to the ocean.

Glaciers

Vast quantities of mud and till are generated by glaciations and deposited on land as till and in lakes. Glaciers can erode already susceptible mudrock formations, and this process enhances glacial production of clay and silt.

The Northern Hemisphere contains 90-percent of the world's lakes larger than 500 km (310 mi), and glaciers created many of those lakes. Lake deposits formed by glaciation, including deep glacial scouring, are abundant.

Non-Glacial Lakes

Although glaciers formed 90-percent of lakes in the Northern Hemisphere, they are not responsible for the formation of ancient lakes. Ancient lakes are the largest and deepest in the world, and hold up to twenty percent of today's petroleum reservoirs. They are also the second most abundant source of mudrocks, behind marine mudrocks.

Ancient lakes owe their abundance of mudrocks to their long lives and thick deposits. These deposits were susceptible to changes in oxygen and rainfall, and offer a robust account of paleoclimate consistency.

Deltas

A delta is a subaerial or subaqueous deposit formed where rivers or streams deposit sediment into a water body. Deltas, such as the Mississippi and Congo, have massive potential for sediment deposit, and can move sediments into deep ocean waters. Delta environments are found at the mouth of a river, where its waters slow as they enter the ocean, and silt and clay are deposited.

The Mississippi Delta.

Low energy deltas, which deposit a great deal of mud, are located in lakes, gulfs, seas, and small oceans, where coastal currents are also low. Sand and gravel-rich deltas are high-energy deltas, where waves dominate, and mud and silt are carried much farther from the mouth of the river.

Coastlines

Coastal currents, mud supply, and waves are a key factor in coastline mud deposition. The Amazon River supplies 500 million tons of sediment, which is mostly clay, to the coastal region of northeastern South America. 250 tons of this sediment moves along the coast and is deposited. Much of the mud accumulated here is more than 20 meters (65 feet) thick, and extends 30 kilometers (19 mi) into the ocean.

Much of the sediment carried by the Amazon can come from the Andes mountains, and the final distance traveled by the sediment is 6,000 km (3,700 mi).

Marine Environments

70-percent of the Earth's surface is covered by ocean, and marine environments are where we find the world's highest proportion of mudrocks. There is a great deal of lateral continuity found in the ocean, as opposed to continents which are confined.

In comparison, continents are temporary stewards of mud and silt, and the inevitable home of mudrock sediments is the oceans. Reference the mudrock cycle below in order to understand the burial and resurgence of the various particles

There are various environments in the oceans, including deep-sea trenches, abyssal plains, volcanic seamounts, convergent, divergent, and transform plate margins. Not only is land a major source of the ocean sediments, but organisms living within the ocean contribute, as well.

The world's rivers transport the largest volume of suspended and dissolved loads of clay and silt to the sea, where they are deposited on ocean shelves. At the poles, glaciers and floating ice drop deposits directly to the sea floor. Winds can provide fine grained material from arid regions, and explosive volcanic eruptions contribute as well. All of these sources vary in the rate of their contribution.

Sediment moves to the deeper parts of the oceans by gravity, and the processes in the ocean are comparable to those on land.

Location has a large impact on the types of mudrocks found in ocean environments. For example, the Apalachicola River, which drains in the subtropics of the United States, carries up to sixty to eighty percent kaolinite mud, whereas the Mississippi carries only ten to twenty percent kaolinite.

The Mudrock Cycle

We can imagine the beginning of a mudrock's life as sediment at the top of a mountain, which may have been uplifted by plate tectonics or propelled into the air from a volcano. This sediment is exposed to rain, wind, and gravity which batters and breaks apart the rock by weathering. The products of weathering, including particles ranging from clay to silt, to pebbles and boulders, are transported to the basin below, where it can solidify into one if its many sedimentary mudstone types.

Eventually, the mudrock will move its way kilometers below the subsurface, where pressure and temperature cook the mudstone into a metamorphosed gneiss. The metamorphosed gneiss will make its way to the surface once again as country rock or as magma in a volcano, and the whole process will begin again.

Important Properties

Color

Mudrocks form in various colors, including: red, purple, brown, yellow, green and grey, and even black. Shades of grey are most common in mudrocks, and darker colors of black come from organic carbons. Green mudrocks form in reducing conditions, where organic matter decomposes along with ferric iron. They can also be found in marine environments, where pelagic, or free-floating species, settle out of the water and decompose in the mudrock. Red mudrocks form when iron within the mudrock becomes oxidized, and depending on the intensity of red, one can determine if the rock has fully oxidized.

Fossils

Fossils are well preserved in mudrock formations, because the fine-grained rock protects the fossils from erosion, dissolution, and other processes of erosion. Fossils are particularly important for recording past environments. Paleontologists can look at a specific area and determine salinity, water depth, water temperature, water turbidity, and sedimentation rates with the aid of type and abundance of fossils in mudrock

Burgess Shale.

One of the most famous mudrock formations is the Burgess Shale in Western Canada, which formed during the Cambrian. At this site, soft bodied creatures were preserved, some in whole, by the activity of mud in a sea. Solid skeletons are, generally, the only remnants of ancient life preserved; however, the Burgess Shale includes hard body parts such as bones, skeletons, teeth, and also soft body parts such as muscles, gills, and digestive systems. The Burgess Shale is one of the most significant fossil locations on Earth, preserving innumerable specimens of 500 million year old species, and its preservation is due to the protection of mudrock.

Another noteworthy formation is the Morrison Formation. This area covers 1.5 million square miles, stretching from Montana to New Mexico in the United States. It is considered one of the world's most significant dinosaur burial grounds, and its many fossils can be found in museums around the world. This site includes dinosaur fossils from a few dinosaur species, including the *Allosaurus*, *Diplodocus*, *Stegosaurus*, and *Brontosaurus*. There are also lungfish, freshwater mollusks, ferns and conifers. This deposit was formed by a humid, tropical climate with lakes, swamps, and rivers, which deposited mudrock. Inevitably, mudrock preserved countless specimens from the late Jurassic, roughly 150 million years ago.

Petroleum and Natural Gas

Mudrocks, especially black shale, are the source and containers of precious petroleum sources throughout the world. Since mudrocks and organic material require quiet water conditions for deposition, mudrocks are the most likely resource for petroleum. Mudrocks have low porosity, they are impermeable, and often, if the mudrock is not black shale, it remains useful as a seal to petroleum and natural gas reservoirs. In the case of petroleum found in a reservoir, the rock surrounding the petroleum is not the source rock, whereas black shale is a source rock.

Why Mudrocks are Important

As noted before, mudrocks make up fifty percent of the Earth's sedimentary geological record. They are widespread on Earth, and important for various industries.

Metamorphosed shale can hold emerald and gold, and mudrocks can host ore metals such as lead and zinc. Mudrocks are important in the preservation of petroleum and natural gas, due to their low porosity, and are commonly used by engineers to inhibit harmful fluid leakage from landfills.

Sandstones and carbonates record high-energy events in our history, and they are much easier to study. Interbedded between the high-energy events are mudrock formations that have recorded quieter, normal conditions in our Earth's history. It is the quieter, normal events of our geologic history we don't yet understand. Sandstones provide the big tectonic picture and some indications of water depth; mudrocks record oxygen content, a generally richer fossil abundance and diversity, and a much more informative geochemistry.

In recognition of mud and mudrocks' sometimes unappreciated importance to earth sciences, the Geological Society of London named 2015 as the "Year of Mud".

Biochemical Sedimentary Rocks

Outcrop of Ordovician oil shale (kukersite), northern Estonia.

Biochemical sedimentary rocks are created when organisms use materials dissolved in air or water to build their tissue. Examples include:

- Most types of limestone are formed from the calcareous skeletons of organisms such as corals, mollusks, and foraminifera.
- Coal, formed from plants that have removed carbon from the atmosphere and combined it with other elements to build their tissue.
- Deposits of chert formed from the accumulation of siliceous skeletons of microscopic organisms such as radiolaria and diatoms.

Chemical Sedimentary Rocks

Chemical sedimentary rock forms when mineral constituents in solution become supersaturated and inorganically precipitate. Common chemical sedimentary rocks include oolitic limestone and rocks composed of evaporite minerals, such as halite (rock salt), sylvite, barite and gypsum.

"Other" Sedimentary Rocks

This fourth miscellaneous category includes rocks formed by Pyroclastic flows, impact breccias, volcanic breccias, and other relatively uncommon processes.

Compositional classification schemes

Alternatively, sedimentary rocks can be subdivided into compositional groups based on their mineralogy:

- Siliciclastic sedimentary rocks, are dominantly composed of silicate minerals. The sediment that makes up these rocks was transported as bed load, suspended load, or by sediment gravity flows. Siliciclastic sedimentary rocks are subdivided into conglomerates and breccias, sandstone, and mudrocks.
- Carbonate sedimentary rocks are composed of calcite (rhombohedral $CaCO_3$), aragonite (orthorhombic $CaCO_3$), dolomite ($CaMg(CO_3)^2$), and other carbonate minerals based on the CO_2^{-3} ion. Common examples include limestone and dolostone.
- Evaporite sedimentary rocks are composed of minerals formed from the evaporation of water. The most common evaporite minerals are carbonates (calcite and others based on CO_2^{-3}), chlorides (halite and others built on Cl−), and sulfates (gypsum and others built on SO_2^{-4}). Evaporite rocks commonly include abundant halite (rock salt), gypsum, and anhydrite.
- Organic-rich sedimentary rocks have significant amounts of organic material, generally in excess of 3% total organic carbon. Common examples include coal, oil shale as well as source rocks for oil and natural gas.
- Siliceous sedimentary rocks are almost entirely composed of silica (SiO_2), typically as chert, opal, chalcedony or other microcrystalline forms.

- Iron-rich sedimentary rocks are composed of >15% iron; the most common forms are banded iron formations and ironstones.
- Phosphatic sedimentary rocks are composed of phosphate minerals and contain more than 6.5% phosphorus; examples include deposits of phosphate nodules, bone beds, and phosphatic mudrocks.

DEPOSITION AND TRANSFORMATION

Sediment Transport and Deposition

Cross-bedding and scour in a fine sandstone; the Logan Formation (Mississippian) of Jackson County, Ohio.

Sedimentary rocks are formed when sediment is deposited out of air, ice, wind, gravity, or water flows carrying the particles in suspension. This sediment is often formed when weathering and erosion break down a rock into loose material in a source area. The material is then transported from the source area to the deposition area. The type of sediment transported depends on the geology of the hinterland (the source area of the sediment). However, some sedimentary rocks, such as evaporites, are composed of material that form at the place of deposition. The nature of a sedimentary rock, therefore, not only depends on the sediment supply, but also on the sedimentary depositional environment in which it formed.

Transformation (Diagenesis)

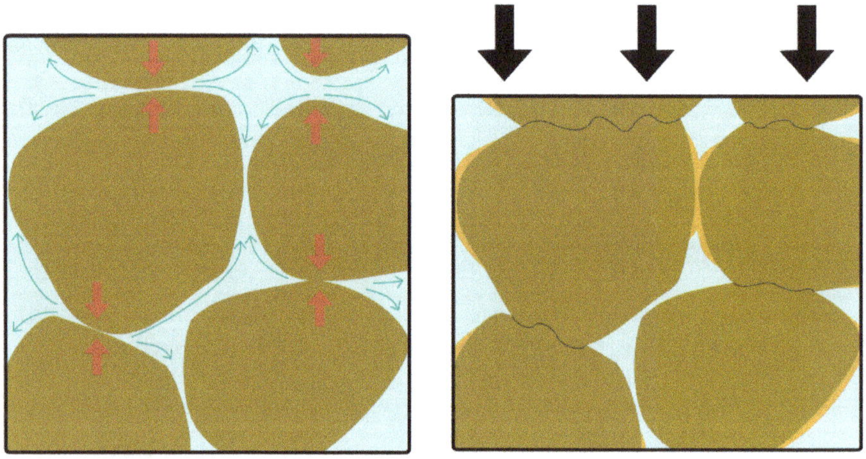

Pressure solution at work in a clastic rock. While material dissolves at places where grains are in contact, that material may recrystallize from the solution and act as cement in open pore spaces. As a result, there is a net flow of material from areas under high stress to those under low stress, producing a sedimentary rock becomes more compact and harder. Loose sand can become sandstone in this way.

The term diagenesis is used to describe all the chemical, physical, and biological changes, exclusive of surface weathering, undergone by a sediment after its initial deposition. Some of those processes cause the sediment to consolidate into a compact, solid substance from the originally loose material. Young sedimentary rocks, especially those of Quaternary age (the most recent period of the geologic time scale) are often still unconsolidated. As sediment deposition builds up, the overburden (lithostatic) pressure rises, and a process known as lithification takes place.

Sedimentary rocks are often saturated with seawater or groundwater, in which minerals can dissolve, or from which minerals can precipitate. Precipitating minerals reduce the pore space in a rock, a process called cementation. Due to the decrease in pore space, the original connate fluids are expelled. The precipitated minerals form a cement and make the rock more compact and competent. In this way, loose clasts in a sedimentary rock can become "glued" together.

When sedimentation continues, an older rock layer becomes buried deeper as a result. The lithostatic pressure in the rock increases due to the weight of the overlying sediment. This causes compaction, a process in which grains mechanically reorganize. Compaction is, for example, an important diagenetic process in clay, which can initially consist of 60% water. During compaction, this interstitial water is pressed out of pore spaces. Compaction can also be the result of dissolution of grains by pressure solution. The dissolved material precipitates again in open pore spaces, which means there is a net flow of material into the pores. However, in some cases, a certain mineral dissolves and does not precipitate again. This process, called leaching, increases pore space in the rock.

Some biochemical processes, like the activity of bacteria, can affect minerals in a rock and are therefore seen as part of diagenesis. Fungi and plants (by their roots) and various other organisms that live beneath the surface can also influence diagenesis.

Burial of rocks due to ongoing sedimentation leads to increased pressure and temperature, which stimulates certain chemical reactions. An example is the reactions by which organic material becomes lignite or coal. When temperature and pressure increase still further, the realm of diagenesis makes way for metamorphism, the process that forms metamorphic rock.

PROPERTIES

A piece of a banded iron formation, a type of rock that consists of alternating layers with iron(III) oxide (red) and iron(II) oxide (grey). BIFs were mostly formed during the Precambrian, when the atmosphere was not yet rich in oxygen. Moories Group, Barberton Greenstone Belt, South Africa.

Color

The color of a sedimentary rock is often mostly determined by iron, an element with two major oxides: iron(II) oxide and iron(III) oxide. Iron(II) oxide (FeO) only forms under low oxygen (anoxic) circumstances and gives the rock a grey or greenish color. Iron(III) oxide (Fe_2O_3) in a richer iron environment is often found in the form of the mineral hematite and gives the rock a reddish to brownish color. In arid continental climates rocks are in direct contact with the atmosphere, and oxidation is an important process, giving the rock a red or orange color. Thick sequences of red sedimentary rocks formed in arid climates are called red beds. However, a red color does not necessarily mean the rock formed in a continental environment or arid climate.

The presence of organic material can color a rock black or grey. Organic material is formed from dead organisms, mostly plants. Normally, such material eventually decays by oxidation or bacterial activity. Under anoxic circumstances, however, organic material cannot decay and leaves a dark sediment, rich in organic material. This can, for example, occur at the bottom of deep seas and lakes. There is little water mixing in such environments, as a result oxygen from surface water is not brought down, and the deposited sediment is normally a fine dark clay. Dark rocks, rich in organic material, are therefore often shales.

Texture

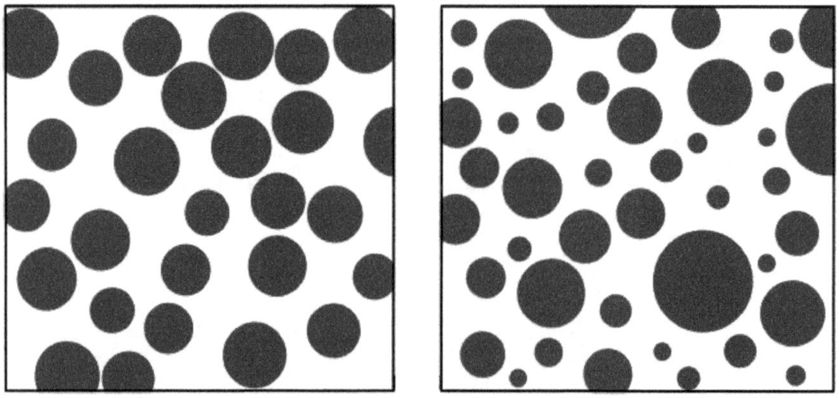

Diagram showing well-sorted (left) and poorly sorted (right) grains.

The size, form and orientation of clasts (the original pieces of rock) in a sediment is called its texture. The texture is a small-scale property of a rock, but determines many of its large-scale properties, such as the density, porosity or permeability.

The 3D orientation of the clasts is called the fabric of the rock. Between the clasts, the rock can be composed of a matrix (a cement) that consists of crystals of one or more precipitated minerals. The size and form of clasts can be used to determine the velocity and direction of current in the sedimentary environment that moved the clasts from their origin; fine, calcareous mud only settles in quiet water while gravel and larger clasts are moved only by rapidly moving water. The grain size of a rock is usually expressed with the Wentworth scale, though alternative scales are sometimes used. The grain size can be expressed as a diameter or a volume, and is always an average value – a rock is composed of clasts with different sizes. The statistical distribution of grain sizes is different for different rock types and is described in a property called the sorting of the rock. When all clasts are more or less of the same size, the rock is called 'well-sorted', and when there is a large spread in grain size, the rock is called 'poorly sorted'.

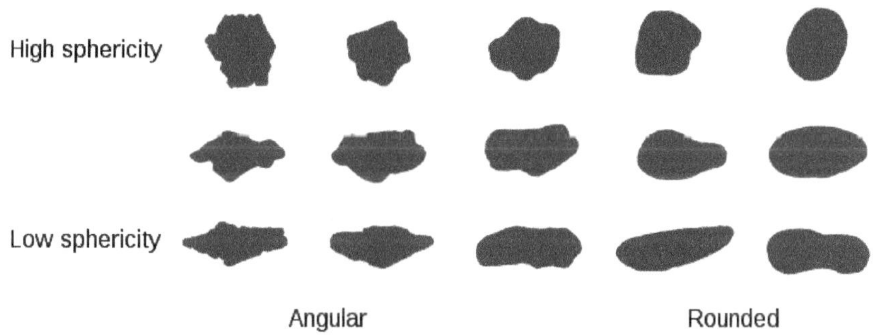

Diagram showing the rounding and sphericity of grains.

The form of the clasts can reflect the origin of the rock.

Coquina, a rock composed of clasts of broken shells, can only form in energetic water. The form of a clast can be described by using four parameters:

- *Surface texture* describes the amount of small-scale relief of the surface of a grain that is too small to influence the general shape.
- *rounding* describes the general smoothness of the shape of a grain.
- 'Sphericity' describes the degree to which the grain approaches a sphere.
- 'Grain form' describes the three dimensional shape of the grain.

Chemical sedimentary rocks have a non-clastic texture, consisting entirely of crystals. To describe such a texture, only the average size of the crystals and the fabric are necessary.

Mineralogy

Most sedimentary rocks contain either quartz (especially siliciclastic rocks) or calcite (especially carbonate rocks). In contrast to igneous and metamorphic rocks, a sedimentary rock usually contains very few different major minerals. However, the origin of the minerals in a sedimentary rock is often more complex than in an igneous rock. Minerals in a sedimentary rock can have formed by precipitation during sedimentation or by diagenesis. In the second case, the mineral precipitate can have grown over an older generation of cement. A complex diagenetic history can be studied by optical mineralogy, using a petrographic microscope.

Carbonate rocks dominantly consist of carbonate minerals such as calcite, aragonite or dolomite. Both the cement and the clasts (including fossils and ooids) of a carbonate sedimentary rock can consist of carbonate minerals. The mineralogy of a clastic rock is determined by the material supplied by the source area, the manner of its transport to the place of deposition and the stability of that particular mineral. The resistance of rock forming minerals to weathering is expressed by Bowen's reaction series. In this series, quartz is the most stable, followed by feldspar, micas, and finally other less stable minerals that are only present when little weathering has occurred. The amount of weathering depends mainly on the distance to the source area, the local climate and the time it took for the sediment to be transported to the point where it is deposited. In most sedimentary rocks, mica, feldspar and less stable minerals have been reduced to clay minerals like kaolinite, illite or smectite.

Fossils

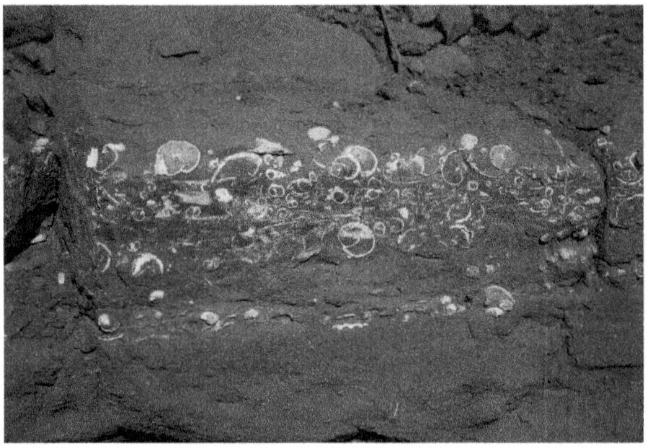

Fossil-rich layers in a sedimentary rock, Año Nuevo State Reserve, California.

Among the three major types of rock, fossils are most commonly found in sedimentary rock. Unlike most igneous and metamorphic rocks, sedimentary rocks form at temperatures and pressures that do not destroy fossil remnants. Often these fossils may only be visible under magnification.

Dead organisms in nature are usually quickly removed by scavengers, bacteria, rotting and erosion, but sedimentation can contribute to exceptional circumstances where these natural processes are unable to work, causing fossilisation. The chance of fossilisation is higher when the sedimentation rate is high (so that a carcass is quickly buried), in anoxic environments (where little bacterial activity occurs) or when the organism had a particularly hard skeleton. Larger, well-preserved fossils are relatively rare.

Fossils can be both the direct remains or imprints of organisms and their skeletons. Most commonly preserved are the harder parts of organisms such as bones, shells, and the woody tissue of plants. Soft tissue has a much smaller chance of being fossilized, and the preservation of soft tissue of animals older than 40 million years is very rare. Imprints of organisms made while they were still alive are called trace fossils, examples of which are burrows, footprints, etc.

As a part of a sedimentary or metamorphic rock, fossils undergo the same diagenetic processes as does the containing rock. A shell consisting of calcite can, for example, dissolve while a cement of silica then fills the cavity. In the same way, precipitating minerals can fill cavities formerly occupied by blood vessels, vascular tissue or other soft tissues. This preserves the form of the organism but changes the chemical composition, a process called permineralization. The most common minerals involved in permineralization are cements of carbonates (especially calcite), forms of amorphous silica (chalcedony, flint, chert) and pyrite. In the case of silica cements, the process is called lithification.

Burrows in a turbidite, made by crustaceans, San Vincente Formation (early Eocene) of the Ainsa Basin, southern foreland of the Pyrenees.

At high pressure and temperature, the organic material of a dead organism undergoes chemical reactions in which volatiles such as water and carbon dioxide are expulsed. The fossil, in the end, consists of a thin layer of pure carbon or its mineralized form, graphite. This form of fossilisation is called carbonisation. It is particularly important for plant fossils. The same process is responsible for the formation of fossil fuels like lignite or coal.

Primary Sedimentary Structures

Cross-bedding in a fluviatilesandstone, Middle Old Red Sandstone (Devonian) on Bressay, Shetland Islands.

A flute cast, a type of sole marking, from the Book Cliffs of Utah.

Ripple marks formed by a current in a sandstone that was later tilted (Haßberge, Bavaria).

Structures in sedimentary rocks can be divided into 'primary' structures (formed during deposition) and 'secondary' structures (formed after deposition). Unlike textures, structures are always large-scale features that can easily be studied in the field. Sedimentary structures can indicate something about the sedimentary environment or can serve to tell which side originally faced up where tectonics have tilted or overturned sedimentary layers.

Sedimentary rocks are laid down in layers called beds or strata. A bed is defined as a layer of rock that has a uniform lithology and texture. Beds form by the deposition of layers of sediment on top of each other. The sequence of beds that characterizes sedimentary rocks is called bedding. Single beds can be a couple of centimeters to several meters thick. Finer, less pronounced layers are called laminae, and the structure it forms in a rock is called lamination. Laminae are usually less than a few centimeters thick. Though bedding and lamination are often originally horizontal in nature, this is not always the case. In some environments, beds are deposited at a (usually small) angle. Sometimes multiple sets of layers with different orientations exist in the same rock, a structure called cross-bedding. Cross-bedding forms when small-scale erosion occurs during deposition, cutting off part of the beds. Newer beds then form at an angle to older ones.

The opposite of cross-bedding is parallel lamination, where all sedimentary layering is parallel. Differences in laminations are generally caused by cyclic changes in the sediment supply, caused, for example, by seasonal changes in rainfall, temperature or biochemical activity. Laminae that represent seasonal changes (similar to tree rings) are called varves. Any sedimentary rock composed of millimeter or finer scale layers can be named with the general term *laminite*. When sedimentary rocks have no lamination at all, their structural character is called massive bedding.

Graded bedding is a structure where beds with a smaller grain size occur on top of beds with larger grains. This structure forms when fast flowing water stops flowing. Larger, heavier clasts in suspension settle first, then smaller clasts. Although graded bedding can form in many different environments, it is a characteristic of turbidity currents.

The surface of a particular bed, called the bedform, can be indicative of a particular sedimentary environment, too. Examples of bed forms include dunes and ripple marks. Sole markings, such as tool marks and flute casts, are groves dug into a sedimentary layer that are preserved. These are often elongated structures and can be used to establish the direction of the flow during deposition.

Ripple marks also form in flowing water. There are two types of ripples: symmetric and asymmetric. Environments where the current is in one direction, such as rivers, produce asymmetric ripples. The longer flank of such ripples is on the upstream side of the current. Symmetric wave ripples occur in environments where currents reverse directions, such as tidal flats.

Mudcracks are a bed form caused by the dehydration of sediment that occasionally comes above the water surface. Such structures are commonly found at tidal flats or point bars along rivers.

Secondary Sedimentary Structures

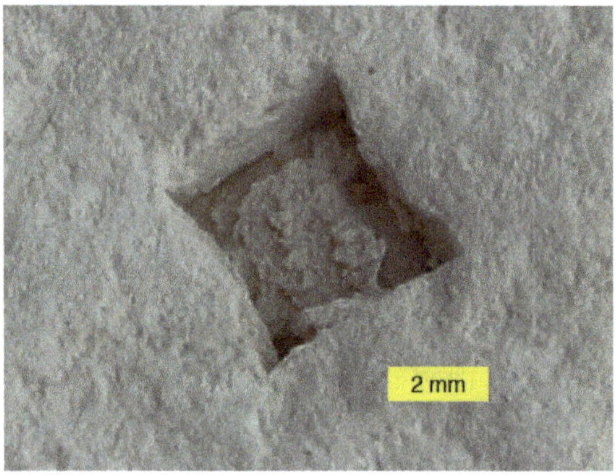

Halite crystal mold in dolomite, Paadla Formation (Silurian), Saaremaa, Estonia.

Secondary sedimentary structures are those which formed after deposition. Such structures form by chemical, physical and biological processes within the sediment. They can be indicators of circumstances after deposition. Some can be used as way up criteria.

Organic materials in a sediment can leave more traces than just fossils. Preserved tracks and burrows are examples of trace fossils (also called ichnofossils). Such traces are relatively rare. Most trace fossils are burrows of molluscs or arthropods. This burrowing is called bioturbation by sedimentologists. It can be a valuable indicator of the biological and ecological environment that existed after the sediment was deposited. On the other hand, the burrowing activity of organisms can destroy other (primary) structures in the sediment, making a reconstruction more difficult.

Chert concretions in chalk, Middle Lefkara Formation (upper Paleocene to middle Eocene), Cyprus.

Secondary structures can also form by diagenesis or the formation of a soil (pedogenesis) when a sediment is exposed above the water level. An example of a diagenetic structure common in carbonate rocks is a stylolite. Stylolites are irregular planes where material was dissolved into the pore fluids in the rock. This can result in the precipitation of a certain chemical species producing coloring and staining of the rock, or the formation of concretions. Concretions are roughly concentric bodies with a different composition from the host rock. Their formation can be the result of localized precipitation due to small differences in composition or porosity of the host rock, such as around fossils, inside burrows or around plant roots. In carbonate based rocks such as limestone or chalk, chert or flint concretions are common, while terrestrial sandstones can have iron concretions. Calcite concretions in clay are called septarian concretions.

After deposition, physical processes can deform the sediment, producing a third class of secondary structures. Density contrasts between different sedimentary layers, such as between sand and clay, can result in flame structures or load casts, formed by inverted diapirism. While the clastic bed is still fluid, diapirism can cause a denser upper layer to sink into a lower layer. Sometimes, density contrasts can result or grow when one of the lithologies dehydrates. Clay can be easily compressed as a result of dehydration, while sand retains the same volume and becomes relatively less dense. On the other hand, when the pore fluid pressure in a sand layer surpasses a critical point, the sand can break through overlying clay layers and flow through, forming discordant bodies of sedimentary rock called sedimentary dykes. The same process can form mud volcanoes on the surface where they broke through upper layers.

Sedimentary dykes can also be formed in a cold climate where the soil is permanently frozen during a large part of the year. Frost weathering can form cracks in the soil that fill with rubble from above. Such structures can be used as climate indicators as well as way up structures.

Density contrasts can also cause small-scale faulting, even while sedimentation progresses (synchronous-sedimentary faulting). Such faulting can also occur when large masses of non-lithified sediment are deposited on a slope, such as at the front side of a delta or the continental slope. Instabilities in such sediments can result in the deposited material to slump, producing fissures and folding. The resulting structures in the rock are syn-sedimentary folds and faults, which can be difficult to distinguish from folds and faults formed by tectonic forces acting on lithified rocks.

SEDIMENTARY ENVIRONMENTS

The setting in which a sedimentary rock forms is called the sedimentary environment. Every environment has a characteristic combination of geologic processes and circumstances. The type of sediment that is deposited is not only dependent on the sediment that is transported to a place, but also on the environment itself.

A marine environment means that the rock was formed in a sea or ocean. Often, a distinction is made between deep and shallow marine environments. Deep marine usually refers to environments more than 200 m below the water surface. Shallow marine environments exist adjacent to coastlines and can extend to the boundaries of the continental shelf. The water movements in such environments have a generally higher energy than that in deep environments, as wave activity diminishes with depth. This means that coarser sediment particles can be transported and the deposited sediment can be coarser than in deeper environments. When the sediment is transported from the continent, an alternation of sand, clay and silt is deposited. When the continent is far away, the amount of such sediment deposited may be small, and biochemical processes dominate the type of rock that forms. Especially in warm climates, shallow marine environments far offshore mainly see deposition of carbonate rocks. The shallow, warm water is an ideal habitat for many small organisms that build carbonate skeletons. When these organisms die, their skeletons sink to the bottom, forming a thick layer of calcareous mud that may lithify into limestone. Warm shallow marine environments also are ideal environments for coral reefs, where the sediment consists mainly of the calcareous skeletons of larger organisms.

In deep marine environments, the water current working the sea bottom is small. Only fine particles can be transported to such places. Typically sediments depositing on the ocean floor are fine clay or small skeletons of micro-organisms. At 4 km depth, the solubility of carbonates increases dramatically (the depth zone where this happens is called the lysocline). Calcareous sediment that sinks below the lysocline dissolves, as a result no limestone can be formed below this depth. Skeletons of micro-organisms formed of silica (such as radiolarians) are not as soluble and still deposit. An example of a rock formed of silica skeletons is radiolarite. When the bottom of the sea has a small inclination, for example at the continental slopes, the sedimentary cover can become unstable, causing turbidity currents. Turbidity currents are sudden disturbances of the normally quite deep marine environment and can cause the geologically speaking instantaneous deposition of large amounts of sediment, such as sand and silt. The rock sequence formed by a turbidity current is called a turbidite.

The coast is an environment dominated by wave action. At a beach, dominantly denser sediment such as sand or gravel, often mingled with shell fragments, is deposited, while the silt and clay sized material is kept in mechanical suspension. Tidal flats and shoals are places that sometimes dry because of the tide. They are often cross-cut by gullies, where the current is strong and the grain size of the deposited sediment is larger. Where rivers enter the body of water, either on a sea or lake coast, deltas can form. These are large accumulations of sediment transported from the continent to places in front of the mouth of the river. Deltas are dominantly composed of clastic sediment (in contrast to chemical).

A sedimentary rock formed on land has a continental sedimentary environment. Examples of continental environments are lagoons, lakes, swamps, floodplains and alluvial fans. In the quiet water of swamps, lakes and lagoons, fine sediment is deposited, mingled with organic material from dead plants and animals. In rivers, the energy of the water is much greater and can transport heavier clastic material. Besides transport by water, sediment can in continental environments also be transported by wind or glaciers. Sediment transported by wind is called aeolian and is always very well sorted, while sediment transported by a glacier is called glacial till and is characterized by very poor sorting.

Aeolian deposits can be quite striking. The depositional environment of the Touchet Formation, located in the Northwestern United States, had intervening periods of aridity which resulted in a series of rhythmite layers. Erosional cracks were later infilled with layers of soil material, especially from aeolian processes. The infilled sections formed vertical inclusions in the horizontally deposited layers of the Touchet Formation, and thus provided evidence of the events that intervened over time among the forty-one layers that were deposited.

Sedimentary Facies

Sedimentary environments usually exist alongside each other in certain natural successions. A beach, where sand and gravel is deposited, is usually bounded by a deeper marine environment a little offshore, where finer sediments are deposited at the same time. Behind the beach, there can be dunes (where the dominant deposition is well sorted sand) or a lagoon (where fine clay and organic material is deposited). Every sedimentary environment has its own characteristic deposits. The typical rock formed in a certain environment is called its sedimentary facies. When sedimentary strata accumulate through time, the environment can shift, forming a change in facies in the subsurface at one location. On the other hand, when a rock layer with a certain age is followed laterally, the lithology (the type of rock) and facies eventually change.

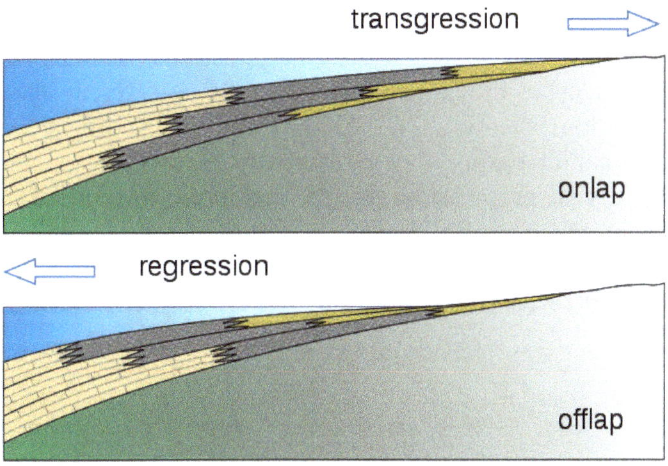

Shifting sedimentary facies in the case of transgression (above) and regression of the sea (below).

Facies can be distinguished in a number of ways: the most common are by the lithology (for example: limestone, siltstone or sandstone) or by fossil content. Coral for example only lives in warm and shallow marine environments and fossils of coral are thus typical for shallow marine facies. Facies determined by lithology are called lithofacies; facies determined by fossils are biofacies.

Sedimentary environments can shift their geographical positions through time. Coastlines can shift in the direction of the sea when the sea level drops, when the surface rises due to tectonic forces in the Earth's crust or when a river forms a large delta. In the subsurface, such geographic shifts of sedimentary environments of the past are recorded in shifts in sedimentary facies. This means that sedimentary facies can change either parallel or perpendicular to an imaginary layer of rock with a fixed age, a phenomenon described by Walther's Law.

The situation in which coastlines move in the direction of the continent is called transgression. In the case of transgression, deeper marine facies are deposited over shallower facies, a succession called onlap. Regression is the situation in which a coastline moves in the direction of the sea. With regression, shallower facies are deposited on top of deeper facies, a situation called offlap.

The facies of all rocks of a certain age can be plotted on a map to give an overview of the palaeogeography. A sequence of maps for different ages can give an insight in the development of the regional geography.

SEDIMENTARY BASINS

Sedimentary basins are regions of Earth of long-term subsidence creating accommodation space for infilling by sediments. The subsidence can result from a variety of causes that include: the thinning of underlying crust, sedimentary, volcanic, and tectonic loading, and changes in the thickness or density of adjacent lithosphere.Sedimentary basins occur in diverse geological settings usually associated with plate tectonic activity. Basins are classified structurally in various ways, with a primary classifications distinguishing among basins formed in various plate tectonic regime (divergent, convergent, transform, intraplate), the proximity of the basin to the active plate margins, and whether oceanic, continental or transitional crust underlies the basin. Basins formed in different plate tectonic regimes vary in their preservation potential. On oceanic crust, basins are likely to be subducted, while marginal continental basins may be partially preserved, and intracratonic basins have a high probability of preservation. As the sediments are buried, they are subjected to increasing pressure and begin the process of lithification. A number of basins formed in extensional settings can undergo inversion which has accounted for a number of the economically viable oil reserves on earth which were formerly basins.

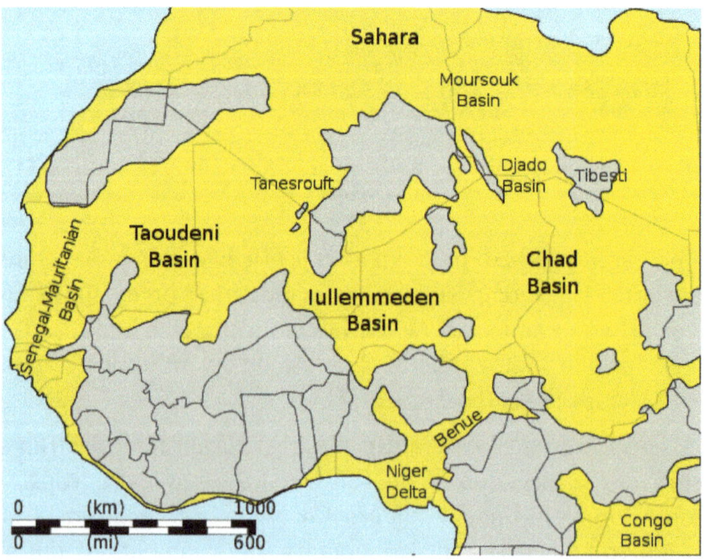

Map of the major sedimentary basins of Central and West Africa.

Methods of Formation

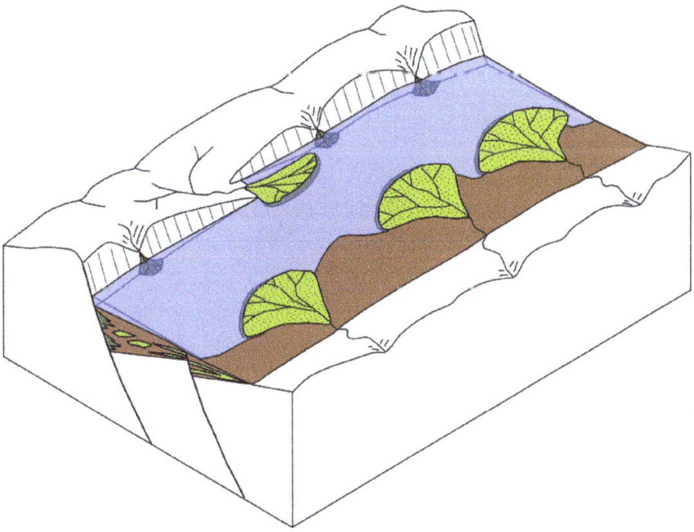

Example of a sedimentary basin in a half-graben.

Sedimentary basins form primarily in convergent, divergent and transform settings. Convergent boundaries create foreland basins through tectonic compression of oceanic and continental crust during lithospheric flexure. Tectonic extension at divergent boundaries where continental rifting is occurring can create a nascent ocean basin leading to either an ocean or the failure of the rift zone. In tectonic strike-slip settings, accommodation spaces occur as transpressional, transtensional or transrotational basins according to the motion of the plates along the fault zone and the local topography pull-apart basins.

Lithospheric Stretching

If the lithosphere is caused to stretch horizontally, by mechanisms such as *ridge-push* or *trench-pull*, the effect is believed to be twofold. The lower, hotter part of the lithosphere will "flow" slowly away from the main area being stretched, whilst the upper, cooler and more brittle crust will tend to fault (crack) and fracture. The combined effect of these two mechanisms is for the Earth's surface in the area of extension to subside, creating a geographical depression which is then often infilled with water and/or sediments. (An analogy might be a piece of rubber, which thins in the middle when stretched.)

An example of a basin caused by lithospheric stretching is the North Sea - also an important location for significant hydrocarbon reserves. Another such feature is the Basin and Range Province which covers most of the USA state of Nevada, forming a series of horst and graben structures.

Another expression of lithospheric stretching results in the formation of ocean basins with central ridges; The Red Sea is in fact an incipient ocean, in a plate tectonic context. The mouth of the Red Sea is also a tectonic triple junction where the Indian Ocean Ridge,

Red Sea Rift and East African Rift meet. This is the only place on the planet where such a triple junction in oceanic crust is exposed subaerially. The reason for this is twofold, due to a high thermal buoyancy of the junction, and a local crumpled zone of seafloor crust acting as a dam against the Red Sea.

Lithospheric Compression/Shortening and Flexure

If a load is placed on the lithosphere, it will tend to flex in the manner of an elastic plate. The magnitude of the lithospheric flexure is a function of the imposed load and the *flexural rigidity* of the lithosphere, and the wavelength of flexure is a function of flexural rigidity alone. Flexural rigidity is in itself, a function of the lithospheric mineral composition, thermal regime, and effective elastic thickness. The nature of the load is varied. For instance, the Hawaiian Islands chain of volcanic edifices has sufficient mass to cause deflection in the lithosphere.

The obduction of one tectonic plate onto another also causes a load and often results in the creation of a foreland basin, such as the Po basin next to the Alps in Italy, the Molasse Basin next to the Alps in Germany, or the Ebro basin next to the Pyrenees in Spain.

Strike-Slip Deformation

Deformation of the lithosphere in the plane of the earth (i.e. such that faults are vertical) occurs as a result of near horizontal maximum and minimum principal stresses. The resulting zones of subsidence are known as strike-slip or pull-apart basins. Basins formed through strike-slip action occur where a vertical fault plane curves. When the curve in the fault plane moves apart, a region of *transtension* results, creating a basin. Another term for a transtensional basin is a *rhombochasm*. A classic rhombochasm is illustrated by the Dead Sea rift, where northward movement of the Arabian Plate relative to the Anatolian Plate has caused a rhombochasm.

The opposite effect is that of *transpression*, where converging movement of a curved fault plane causes collision of the opposing sides of the fault. An example is the San Bernardino Mountains north of Los Angeles, which result from convergence along a curve in the San Andreas fault system. The Northridge earthquake was caused by vertical movement along local thrust and reverse faults *bunching up* against the bend in the otherwise strike-slip fault environment. In Nigeria, the dominant type of basement rock intersected by wells drilled for hydrocarbons, limestone, or water is granite. The three sedimentary basins in Nigeria are underlain by continental crust except in the Niger delta, where the basement rock is interpreted to be oceanic crust. Most of the wells that penetrated the basement are in the Eastern Dahomey embayment of western Nigeria. A maximum thickness of about 12,000 m of sedimentary rocks is attained in the offshore western Niger delta, but maximum thicknesses of sedimentary rocks are about 2,000 m in the Chad basin and only 500 m in the Sokoto embayment.

Ongoing Development

As more and more sediment is deposited into the basin, the weight of all the newer sediment may cause the basin to subside further because of isostasy. A basin can continue having sediment deposited into it, and continue to subside, for long periods of geological time; this can result in basins many kilometres in thickness. Geologic faults can often occur around the edge of, and within, the basin, as a result of the ongoing slippage and subsidence.

Study of Sedimentary Basins

The study of sedimentary basins as a specific entity in themselves is often referred to as basin modelling or sedimentary basin analysis. The need to understand the processes of basin formation and evolution are not restricted to the purely academic. Indeed, sedimentary basins are the location for almost all of the world's hydrocarbon reserves and as such are the focus of intense commercial interest.

Influence of Astronomical Cycles

In many cases facies changes and other lithological features in sequences of sedimentary rock have a cyclic nature. This cyclic nature was caused by cyclic changes in sediment supply and the sedimentary environment. Most of these cyclic changes are caused by astronomic cycles. Short astronomic cycles can be the difference between the tides or the spring tide every two weeks. On a larger time-scale, cyclic changes in climate and sea level are caused by Milankovitch cycles: cyclic changes in the orientation and/or position of the Earth's rotational axis and orbit around the Sun. There are a number of Milankovitch cycles known, lasting between 10,000 and 200,000 years.

Relatively small changes in the orientation of the Earth's axis or length of the seasons can be a major influence on the Earth's climate. An example are the ice ages of the past 2.6 million years (the Quaternary period), which are assumed to have been caused by astronomic cycles. Climate change can influence the global sea level (and thus the amount of accommodation space in sedimentary basins) and sediment supply from a certain region. Eventually, small changes in astronomic parameters can cause large changes in sedimentary environment and sedimentation.

Sedimentation Rates

The rate at which sediment is deposited differs depending on the location. A channel in a tidal flat can see the deposition of a few metres of sediment in one day, while on the deep ocean floor each year only a few millimetres of sediment accumulate. A distinction can be made between normal sedimentation and sedimentation caused by catastrophic processes. The latter category includes all kinds of sudden exceptional processes like mass movements, rock slides or flooding. Catastrophic processes can see the sudden deposition of a large amount of sediment at once. In some sedimentary environments, most of the total column of sedimentary rock was formed by catastrophic processes, even though the environment is usually a quiet place. Other sedimentary environments are dominated by normal, ongoing sedimentation.

In many cases, sedimentation occurs slowly. In a desert, for example, the wind deposits siliciclastic material (sand or silt) in some spots, or catastrophic flooding of a wadi may cause sudden deposits of large quantities of detrital material, but in most places eolian erosion dominates. The amount of sedimentary rock that forms is not only dependent on the amount of supplied material, but also on how well the material consolidates. Erosion removes most deposited sediment shortly after deposition.

Stratigraphy

The Permian through Jurassic stratigraphy of the Colorado Plateau area of southeastern Utah that makes up much of the famous prominent rock formations in protected areas such as Capitol Reef National Park and Canyonlands National Park. From top to bottom: Rounded tan domes of the Navajo Sandstone, layered red Kayenta Formation, cliff-forming, vertically jointed, red Wingate Sandstone, slope-forming, purplish Chinle Formation, layered, lighter-red Moenkopi Formation, and white, layered Cutler Formation sandstone. Picture from Glen Canyon National Recreation Area, Utah.

That new rock layers are above older rock layers is stated in the principle of superposition. There are usually some gaps in the sequence called unconformities. These represent periods where no new sediments were laid down, or when earlier sedimentary layers were raised above sea level and eroded away.

Sedimentary rocks contain important information about the history of the Earth. They contain fossils, the preserved remains of ancient plants and animals. Coal is considered a type of sedimentary rock. The composition of sediments provides us with clues as to the original rock. Differences between successive layers indicate changes to the environment over time. Sedimentary rocks can contain fossils because, unlike most igneous and metamorphic rocks, they form at temperatures and pressures that do not destroy fossil remains.

REFERENCES

1. Wilkinson, Bruce H.; McElroy, Brandon J.; Kesler, Stephen E.; Peters, Shanan E.; Rothman, Edward D. (2008). "Global geologic maps are tectonic speedometers—Rates of rock cycling from area-age frequencies". Geological Society of America Bulletin. 121: 760–779. doi:10.1130/B26457.1.
2. Buchner & Grapes (2011), p. 24
3. Dott (1964)
4. Blatt et al. (1980), p. 782
5. Prothero & Schwab (2004)
6. Boggs (2006)
7. Stow (2005)
8. Levin (1987), p. 57
9. Tarbuck & Lutgens (1999), pp. 145–146
10. Boggs (1987), p. 105
11. Tarbuck & Lutgens (1999), pp. 156–157
12. Levin (1987), p. 58
13. Boggs (1987), pp. 112–115
14. Blatt et al. (1980), pp. 55–58
15. Levin (1987), p. 60
16. Blatt et al. (1980), pp. 75–80
17. Folk (1965), p. 62
18. For an overview of major minerals in siliciclastic rocks and their relative stabilities, see Folk (1965), pp. 62–64.
19. Stanley (1999), pp. 60–61
20. Levin (1987), p. 92
21. Stanley (1999), p. 61
22. Levin (1987), pp. 92–93
23. Tarbuck & Lutgens (1999), pp. 160–161
24. Press et al. (2003), p. 171
25. Boggs (1987), p. 138
26. For descriptions of cross-bedding, see Blatt et al. (1980), p. 128, pp. 135–136; Press et al. (2003), pp. 171–172.
27. Blatt et al. (1980), pp. 133–135
28. For an explanation about graded bedding, see Boggs (1987), pp. 143–144; Tarbuck & Lutgens (1999), p. 161; Press et al. (2003), p. 172.
29. Collinson et al. (2006), pp. 46–52
30. Blatt et al. (1980), pp. 155–157

31. Tarbuck & Lutgens (1999), p. 162
32. Levin (1987), p. 62
33. Blatt et al. (1980), pp. 136–154
34. For a short description of trace fossils, see Stanley (1999), p. 62; Levin (1987), pp. 93–95; and Collinson et al. (2006), pp. 216–232.
35. Collinson et al. (2006), p. 215
36. For concretions, see Collinson et al. (2006), pp. 206–215.
37. Collinson et al. (2006), pp. 183–185
38. Collinson et al. (2006), pp. 193–194
39. Collinson et al. (2006), pp. 202–203
40. For an overview of different sedimentary environments, see Press et al. (2003) or Einsele (2000), part II.
41. For a definition of shallow marine environments, see Levin (2003), p. 63.
42. Tarbuck & Lutgens (1999), pp. 452–453
43. For an overview of continental environments, see Levin (2003), pp. 67–68.
44. Baker, Victor R.; Nummedal, Dag, eds. (1978). The Channeled Scabland: A Guide to the Geomorphology of the Columbia Basin, Washington. Washington, D.C.: Planetary Geology Program, Office of Space Science, National Aeoronautics and Space Administration. pp. 173–177. ISBN 0-88192-590-X.
45. Tarbuck & Lutgens (1999), pp. 158–160
46. Reading (1996), pp. 19–20
47. Reading (1996), pp. 20–21
48. For an overview over facies shifts and the relations in the sedimentary rock record by which they can be recognized, see Reading (1996), pp. 22–33.
49. For an overview of sedimentary basin types, see Press et al. (2003), pp. 187–189; Einsele (2000), pp. 3–9.
50. For a short explanation of Milankovitch cycles, see Tarbuck & Lutgens (1999), pp. 322–323; Reading (1996), pp. 14–15.
51. Stanley (1999), p. 536
52. Andersen & Borns (1994), pp. 29–32
53. Reading (1996), p. 17
54. Essentials of Geology, 3rd Ed, Stephen Marshak, p G-3
55. Essentials of Geology, 3rd Ed, Stephen Marshak, p G-5
56. Boggs, Jr., Sam. Principles of Sedimentology and Stratigraphy. Pearson Prentice Hall: Upper Saddle River, New Jersey, 2006
57. Neuendorf, Klaus; Mehl, James; Jackson, Julia Glossary of Geology, Fifth Edition. American Geological Institute: Alexandria, VA; 2005.
58. Dott, R. H., Wacke, graywacke and matrix - What Approach to Immature Sandstone Classification: Journal of Sedimentary Petrology, v-34, p. 625-632.,

1996.

59. Spears, D.A., Sam. Towards a classification of Shales. J. geol. soc., London, 137, 1990.

60. Blatt, h., Middleton, G. V. & Murray, R. C. 1972. Origin of Sedimentary Rocks. Prentice Hall Inc., Englewood Cliffs, 634 pp.

61. Boggs, S. (2006) Principles of Sedimentology and Stratigraphy., 2nd ed. Printice Hall, New York. 662 pp. ISBN 0-13-154728-3

62. Friedman, G.M. (2003) Classification of sediments and sedimentary rocks. In Gerard V. Middleton, ed., pp. 127-135, Encyclopedia of Sediments & Sedimentary Rocks, Encyclopedia of Earth Science Series. Kluwer Academic Publishers, Boston, Massachusetts. 821 pp. ISBN 978-1-4020-0872-6

63. Neuendorf, K.K.E., J.P. Mehl, Jr., and J.A. Jackson, eds. (2005) Glossary of Geology (5th ed.). Alexandria, Virginia, American Geological Institute. 779 pp. ISBN 0-922152-76-4

64. Nichols, G. (2009) Sedimentology and Stratigraphy, 2nd ed. John Wiley & Sons Ltd, Chichester, West Sussex, United Kingdom. 419 pp. ISBN 978-1-4051-9379-5

65. Tucker, M.E. (2003) Sedimentary Rocks in the Field, 3rd ed. John Wiley & Sons Ltd,West Sussex, England. 234 pp. ISBN 0-470-85123-6

66. Williams, G.D. (1966) Origin of Shale-Pebble Conglomerate. American Association of Petroleum Geologist Bulletin. vol. 50, no. 3, pp. 573-577.

67. Flugel, E. (2010) Microfacies of Carbonate Rocks: Analysis, Interpretation and Application, 2nd ed. Springer-Verlag, Berlin, Germany. 984 pp. ISBN 978-3-642-03795-5

68. Walker, R.G. 1979. Facies Models. Reprinted with revisions from a series of papers in Geoscience Canada, 1976–1979, Geological Association of Canada, 211pp.

69. Seibold, E. & Berger, W.H. 1996. The Sea Floor: an introduction to Marine Geology, Springer, 356pp.

70. Tucker, M.E. 2001. Sedimentary petrology, 3rd edition, WileyBlackwell, 272pp.

71. "Geologic map of Schuylkill County". 1884.

72. "NASA Rover Finds Old Streambed on Martian Surface". News. Jet Propulsion Laboratory / California Institute of Technology. 27 September 2012. Retrieved 28 February 2016.

CHAPTER 4

Sediment Transport

Chapter Contents

Sediment transport is the movement of organic and inorganic particles by water. In general, the greater the flow, the more sediment that will be conveyed. Water flow can be strong enough to suspend particles in the water column as they move downstream, or simply push them along the bottom of a waterway. Transported sediment may include mineral matter, chemicals and pollutants, and organic material.

Another name for sediment transport is sediment load. The total load includes all particles moving as bedload, suspended load, and wash load.

INTRODUCTION

Sediment transport is the movement of solid particles (sediment), typically due to a combination of gravity acting on the sediment, and/or the movement of the fluid in which the sediment is entrained. Sediment transport occurs in natural systems where the particles are clastic rocks (sand, gravel, boulders, etc.), mud, or clay; the fluid is air, water, or ice; and the force of gravity acts to move the particles along the sloping surface on which they are resting. Sediment transport due to fluid motion occurs in rivers, oceans, lakes, seas, and other bodies of water due to currents and tides. Transport is also caused by glaciers as they flow, and on terrestrial surfaces under the influence of wind. Sediment transport due only to gravity can occur on sloping surfaces in general, including hillslopes, scarps, cliffs, and the continental shelf—continental slope boundary.

Sediment transport is important in the fields of sedimentary geology, geomorphology, civil engineering and environmental engineering (see applications, below). Knowledge of sediment transport is most often used to determine whether erosion or deposition will occur, the magnitude of this erosion or deposition, and the time and distance over which it will occur.

Dust blows from the Sahara Desert over the Atlantic Ocean towards the Canary Islands.

MECHANISMS

Sand blowing off a crest in the Kelso Dunes of the Mojave Desert, California.

Toklat River, East Fork, Polychrome overlook, Denali National Park, Alaska. This river, like other braided streams, rapidly changes the positions of its channels through processes of erosion, sediment transport, and deposition.

Congo river viewed from Kinshasa, Democratic Republic of Congo. Its brownish color is mainly the result of the transported sediments taken upstream.

Aeolian

Aeolian or *eolian* (depending on the parsing of æ) is the term for sediment transport by wind. This process results in the formation of ripples and sand dunes. Typically, the size of the transported sediment is fine sand (<1 mm) and smaller, because air is a fluid with low density and viscosity, and can therefore not exert very much shear on its bed.

Bedforms are generated by aeolian sediment transport in the terrestrial near-surface environment. Ripples and dunes form as a natural self-organizing response to sediment transport.

Aeolian sediment transport is common on beaches and in the arid regions of the world, because it is in these environments that vegetation does not prevent the presence and motion of fields of sand.

Wind-blown very fine-grained dust is capable of entering the upper atmosphere and moving across the globe. Dust from the Sahara deposits on the Canary Islands and islands in the Caribbean, and dust from the Gobi desert has deposited on the western United States. This sediment is important to the soil budget and ecology of several islands.

Deposits of fine-grained wind-blown glacial sediment are called loess.

Fluvial

In geology, physical geography, and sediment transport, fluvial processes relate to flowing water in natural systems. This encompasses rivers, streams, periglacial flows, flash floods and glacial lake outburst floods. Sediment moved by water can be larger than sediment moved by air because water has both a higher density and viscosity. In typical rivers the largest carried sediment is of sand and gravel size, but larger floods can carry cobbles and even boulders.

Fluvial sediment transport can result in the formation of ripples and dunes, in fractal-shaped patterns of erosion, in complex patterns of natural river systems, and in the development of floodplains.

Sand ripples, Laysan Beach, Hawaii. Coastal sediment transport results in these evenly spaced ripples along the shore. Monk seal for scale.

Coastal

Coastal sediment transport takes place in near-shore environments due to the motions of waves and currents. At the mouths of rivers, coastal sediment and fluvial sediment transport processes mesh to create river deltas.

Coastal sediment transport results in the formation of characteristic coastal landforms such as beaches, barrier islands, and capes.

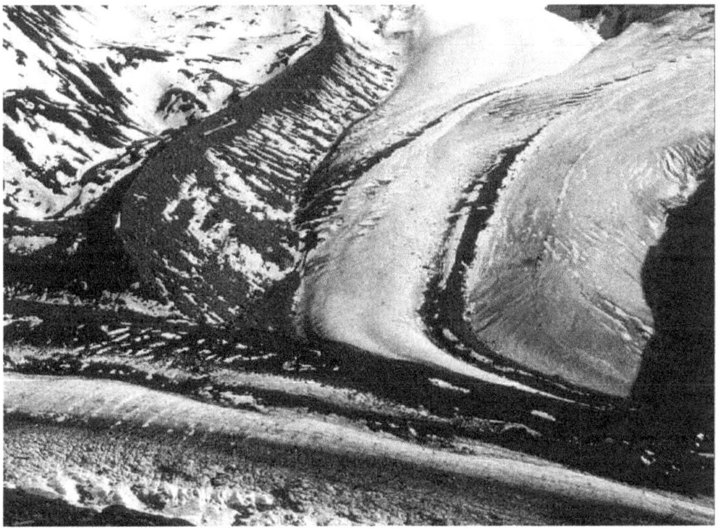

A glacier joining the Gorner Glacier, Zermatt, Switzerland. These glaciers transport sediment and leave behind lateral moraines.

Glacial

As glaciers move over their beds, they entrain and move material of all sizes. Glaciers can carry the largest sediment, and areas of glacial deposition often contain a large number of glacial erratics, many of which are several meters in diameter. Glaciers also pulverize rock into "glacial flour", which is so fine that it is often carried away by winds to create loess deposits thousands of kilometers afield. Sediment entrained in glaciers often moves approximately along the glacial flowlines, causing it to appear at the surface in the ablation zone.

Hillslope

In hillslope sediment transport, a variety of processes move regolith downslope. These include:

- Soil creep
- Tree throw

- Movement of soil by burrowing animals
- Slumping and landsliding of the hillslope

These processes generally combine to give the hillslope a profile that looks like a solution to the diffusion equation, where the diffusivity is a parameter that relates to the ease of sediment transport on the particular hillslope. For this reason, the tops of hills generally have a parabolic concave-up profile, which grades into a convex-up profile around valleys.

As hillslopes steepen, however, they become more prone to episodic landslides and other mass wasting events. Therefore, hillslope processes are better described by a nonlinear diffusion equation in which classic diffusion dominates for shallow slopes and erosion rates go to infinity as the hillslope reaches a critical angle of repose.

Debris Flow

Large masses of material are moved in debris flows, hyperconcentrated mixtures of mud, clasts that range up to boulder-size, and water. Debris flows move as granular flows down steep mountain valleys and washes. Because they transport sediment as a granular mixture, their transport mechanisms and capacities scale differently from those of fluvial systems.

APPLICATIONS

Sediment transport is applied to solve many environmental, geotechnical, and geological problems. Measuring or quantifying sediment transport or erosion is therefore important for coastal engineering. Several sediment erosion devices have been designed in order to quantitfy sediment erosion (e.g., Particle Erosion Simulator (PES)). One such device, also referred to as the BEAST (Benthic Environmental Assessment Sediment Tool) has been calibrated in order to quantify rates of sediment erosion.

Suspended sediment from a stream emptying into a fjord (Isfjorden, Svalbard, Norway).

Movement of sediment is important in providing habitat for fish and other organisms in rivers. Therefore, managers of highly regulated rivers, which are often sediment-starved due to dams, are often advised to stage short floods to refresh the bed material and rebuild bars. This is also important, for example, in the Grand Canyon of the Colorado River, to rebuild shoreline habitats also used as campsites.

Sediment discharge into a reservoir formed by a dam forms a reservoir delta. This delta will fill the basin, and eventually, either the reservoir will need to be dredged or the dam will need to be removed. Knowledge of sediment transport can be used to properly plan to extend the life of a dam.

Geologists can use inverse solutions of transport relationships to understand flow depth, velocity, and direction, from sedimentary rocks and young deposits of alluvial materials.

Flow in culverts, over dams, and around bridge piers can cause erosion of the bed. This erosion can damage the environment and expose or unsettle the foundations of the structure. Therefore, good knowledge of the mechanics of sediment transport in a built environment are important for civil and hydraulic engineers.

When suspended sediment transport is increased due to human activities, causing environmental problems including the filling of channels, it is called siltation after the grain-size fraction dominating the process.

INITIATION OF MOTION

Stress Balance

For a fluid to begin transporting sediment that is currently at rest on a surface, the boundary (or bed) shear stress τ_b exerted by the fluid must exceed the critical shear stress τ_c for the initiation of motion of grains at the bed. This basic criterion for the initiation of motion can be written as:

$$\tau_b = \tau_c .$$

This is typically represented by a comparison between a dimensionless shear stress $(\tau_b *)$and a dimensionless critical shear stress $(\tau_c *)$. The nondimensionalization is in order to compare the driving forces of particle motion (shear stress) to the resisting forces that would make it stationary (particle density and size). This dimensionless shear stress, $\tau *$, is called the Shields parameter and is defined as:

$$\tau * = \frac{\tau}{(\rho_s - \rho_f)(g)(D)} .$$

And the new equation to solve becomes:

$$\tau_b * = \tau_c * .$$

The equations included here describe sediment transport for clastic, or granular sediment. They do not work for clays and muds because these types of floccular sediments do not fit the geometric simplifications in these equations, and also interact thorough electrostatic forces. The equations were also designed for fluvial sediment transport of particles carried along in a liquid flow, such as that in a river, canal, or other open channel.

Only one size of particle is considered in this equation. However, river beds are often formed by a mixture of sediment of various sizes. In case of partial motion where only a part of the sediment mixture moves, the river bed becomes enriched in large gravel as the smaller sediments are washed away. The smaller sediments present under this layer of large gravel have a lower possibility of movement and total sediment transport decreases. This is called armoring effect. Other forms of armoring of sediment or decreasing rates of sediment erosion can be caused by carpets of microbial mats, under conditions of high organic loading.

Critical Shear Stress

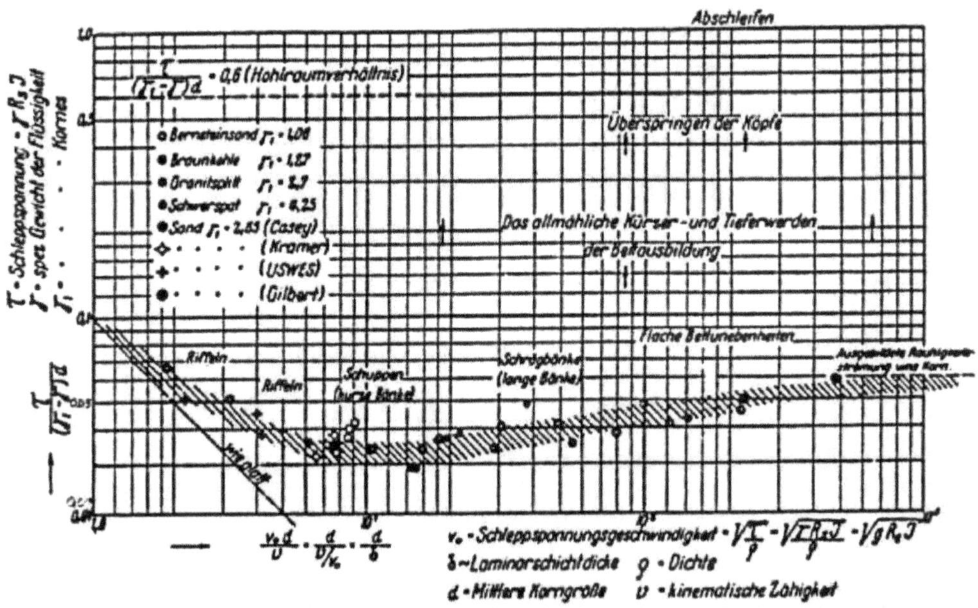

Original Shields diagram, 1936.

The Shields diagram empirically shows how the dimensionless critical shear stress (i.e. the dimensionless shear stress required for the initiation of motion) is a function of a particular form of the particle Reynolds number, Re_p or Reynolds number related to the particle. This allows us to rewrite the criterion for the initiation of motion in terms of only needing to solve for a specific version of the particle Reynolds number, which we call $\mathrm{Re}_p *$.

$$\tau_b * = f\left(\mathrm{Re}_p *\right)$$

This equation can then be solved by using the empirically derived Shields curve to find $\mathrm{Re}_p *$ as a function of a specific form of the particle Reynolds number called the boundary Reynolds number. The mathematical solution of the equation was given by Dey.

Particle Reynolds Number

In general, a particle Reynolds Number has the form:

$$\mathrm{Re}_p = \frac{U_p D}{\nu}$$

Where U_p is a characteristic particle velocity, D is the grain diameter (a characteristic particle size), and ν is the kinematic viscosity, which is given by the dynamic viscosity, μ, divided by the fluid density, ρ_f.

$$\nu = \frac{\mu}{\rho_f}$$

The specific particle Reynolds number of interest is called the boundary Reynolds number, and it is formed by replacing the velocity term in the Particle Reynolds number by the shear velocity, u_*, which is a way of rewriting shear stress in terms of velocity.

$$u_* = \sqrt{\frac{\tau_b}{\rho_f}} = \kappa z \frac{\partial u}{\partial z}$$

where τ_b is the bed shear stress (described below), and κ is the von Kármán constant, where

$$\kappa = 0.407.$$

The particle Reynolds number is therefore given by:

$$\mathrm{Re}_p * = \frac{u_* D}{\nu}$$

Bed Shear Stress

The boundary Reynolds number can be used with the Shields diagram to empirically solve the equation

$$\tau_c * = f(\mathrm{Re}_p *),$$

which solves the right-hand side of the equation

$$\tau_b * = \tau_c *.$$

In order to solve the left-hand side, expanded as

$$\tau_b * = \frac{\tau_b}{(\rho_s - \rho_f)(g)(D)},$$

we must find the bed shear stress, τ_b. There are several ways to solve for the bed shear stress. First, we develop the simplest approach, in which the flow is assumed to be steady and uniform and reach-averaged depth and slope are used. Due to the difficulty of measuring shear stress *in situ*, this method is also one of the most-commonly used. This method is known as the depth-slope product.

Depth-Slope Product

For a river undergoing approximately steady, uniform equilibrium flow, of approximately constant depth h and slope angle θ over the reach of interest, and whose width is much greater than its depth, the bed shear stress is given by some momentum considerations stating that the gravity force component in the flow direction equals exactly the friction force. For a wide channel, it yields:

$$\tau_b = \rho g h \sin(\theta)$$

For shallow slope angles, which are found in almost all natural lowland streams, the small-angle formula shows that $\sin(\theta)$ is approximately equal to $\tan(\theta)$, which is given by S, the slope. Rewritten with this:

$$\tau_b = \rho g h S$$

Shear Velocity, Velocity, and Friction Factor

For the steady case, by extrapolating the depth-slope product and the equation for shear velocity:

$$\tau_b = \rho g h S$$

$$u_* = \sqrt{\left(\frac{\tau_b}{\rho}\right)},$$

We can see that the depth-slope product can be rewritten as:

$$\tau_b = \rho u_*^2.$$

$u*$ is related to the mean flow velocity, $\bar{u}$, through the generalized Darcy-Weisbach friction factor, C_f, which is equal to the Darcy-Weisbach friction factor divided by 8 (for mathematical convenience). Inserting this friction factor,

$$\tau_b = \rho C_f (\bar{u})^2.$$

Unsteady Flow

For all flows that cannot be simplified as a single-slope infinite channel (as in the depth-slope product, above), the bed shear stress can be locally found by applying the Saint-Vennant equations for continuity, which consider accelerations within the flow.

Example

Set-Up

The criterion for the initiation of motion, established earlier, states that

$$\tau_b* = \tau_c *.$$

In this equation,

$$\tau* = \frac{\tau}{(\rho_s - \rho)(g)(D)}, \text{ and therefore}$$

$$\frac{\tau_b}{(\rho_s - \rho)(g)(D)} = \frac{\tau_c}{(\rho_s - \rho)(g)(D)}.$$

$\tau_c *$ is a function of boundary Reynolds number, a specific type of particle Reynolds number.

$$\tau_c* = f\,(Re_p*).$$

For a particular particle Reynolds number, $\tau_c *$ will be an emprical constant given by the Shields Curve or by another set of empirical data (depending on whether or not the grain size is uniform).

Therefore, the final equation that we seek to solve is:

$$\frac{\tau_b}{(\rho_s - \rho)(g)(D)} = f\,(Re_p*).$$

Solution

We make several assumptions to provide an example that will allow us to bring the above form of the equation into a solved form.

First, we assume that the a good approximation of reach-averaged shear stress is given by the depth-slope product. We can then rewrite the equation as

$$\rho ghS = 0.06(\rho_s - \rho)(g)(D).$$

Moving and re-combining the terms, we obtain:

$$hS = \frac{(\rho_s - \rho)}{\rho}(D)\,(f\,(\text{Re}_p*)) = RD\,(f\,(\text{Re}_p*))$$

where R is the submerged specific gravity of the sediment.

We then make our second assumption, which is that the particle Reynolds number is high. This is typically applicable to particles of gravel-size or larger in a stream, and means that the critical shear stress is a constant. The Shields curve shows that for a bed with a uniform grain size,

$$\tau_c* = 0.06.$$

Later researchers have shown that this value is closer to

$$\tau_c* = 0.03$$

for more uniformly sorted beds. Therefore, we will simply insert

$$\tau_c* = f\,(\mathrm{Re}_p*)$$

and insert both values at the end.

The equation now reads:

$$hS = RD\tau_c\,*$$

This final expression shows that the product of the channel depth and slope is equal to the Shield's criterion times the submerged specific gravity of the particles times the particle diameter.

For a typical situation, such as quartz-rich sediment $\left(\rho_s = 2650\frac{kg}{m^3}\right)$ in water $\left(\rho = 1000\frac{kg}{m^3}\right)$, the submerged specific gravity is equal to 1.65.

$$R = \frac{(\rho_s - \rho)}{\rho} = 1.65$$

Plugging this into the equation above,

$$hS = 1.65(D)\tau_c\,*.$$

For the Shield's criterion of $\tau_c* = 0.06$. 0.06 * 1.65 = 0.099, which is well within standard margins of error of 0.1. Therefore, for a uniform bed,

$$hS = 0.1(D).$$

For these situations, the product of the depth and slope of the flow should be 10% of the diameter of the median grain diameter.

The mixed-grain-size bed value is $\tau_c* = 0.03$, which is supported by more recent research as being more broadly applicable because most natural streams have mixed grain sizes. Using this value, and changing D to D_50 ("50" for the 50th percentile, or the median grain size, as we are now looking at a mixed-grain-size bed), the equation becomes:

$$hS = 0.05(D_{50})$$

Which means that the depth times the slope should be about 5% of the median grain diameter in the case of a mixed-grain-size bed.

MODES OF ENTRAINMENT

The sediments entrained in a flow can be transported along the bed as bed load in the form of sliding and rolling grains, or in suspension as suspended load advected by the main flow.Some sediment materials may also come from the upstream reaches and be carried downstream in the form of wash load.

Rouse Number

The location in the flow in which a particle is entrained is determined by the Rouse number, which is determined by the density ρ_s and diameter d of the sediment particle, and the density ρ and kinematic viscosity v of the fluid, determine in which part of the flow the sediment particle will be carried.

$$P = \frac{w_s}{\kappa u_*}$$

Here, the Rouse number is given by P. The term in the numerator is the (downwards) sediment the sediment settling velocity w_s, which is discussed below. The upwards velocity on the grain is given as a product of the von Kármán constant, $\kappa = 0.4$, and the shear velocity, u_*.

The following table gives the approximate required Rouse numbers for transport as bed load, suspended load, and wash load.

Mode of Transport	Rouse Number
Initiation of motion	>7.5
Bed load	>2.5, <7.5
Suspended load: 50% Suspended	>1.2, <2.5
Suspended load: 100% Suspended	>0.8, <1.2
Wash load	<0.8

Settling Velocity

The settling velocity (also called the "fall velocity" or "terminal velocity") is a function of the particle Reynolds number. Generally, for small particles (laminar approximation), it can be calculated with Stokes' Law. For larger particles (turbulent particle Reynolds numbers), fall velocity is calculated with the turbulent drag law. Dietrich (1982) compiled a large amount of published data to which he empirically fit settling velocity curves. Ferguson and Church (2006) analytically combined the expressions for Stokes flow and a turbulent drag law into a single equation that works for all sizes of sediment, and successfully tested it against the data of Dietrich. Their equation is

$$w_s = \frac{RgD^2}{C_1 \nu + (0.75 C_2 RgD^3)^{(0.5)}}.$$

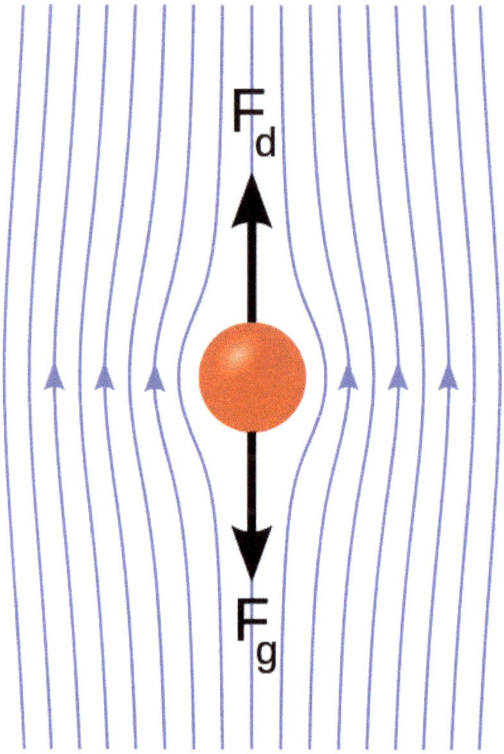

Streamlines around a sphere falling through a fluid. This illustration is accurate for laminar flow, in which the particle Reynolds number is small. This is typical for small particles falling through a viscous fluid; larger particles would result in the creation of a turbulent wake.

In this equation w_s is the sediment settling velocity, g is acceleration due to gravity, and D is mean sediment diameter. ν is the kinematic viscosity of water, which is approximately 1.0 x 10^{-6} m^2/s for water at 20 °C.

C_1 and C_2 are constants related to the shape and smoothness of the grains.

Constant	Smooth Spheres	Natural Grains: Sieve Diameters	Natural Grains: Nominal Diameters	Limit for Ultra-Angular Grains
{\displaystyle C_{1}}	18	18	20	24
{\displaystyle C_{2}}	0.4	1.0	1.1	1.2

The expression for fall velocity can be simplified so that it can be solved only in terms of D. We use the sieve diameters for natural grains, $g = 9.8$, and values given above for ν and R. From these parameters, the fall velocity is given by the expression:

$$w_s = \frac{16.17D^2}{1.8 \cdot 10^{-5} + (12.1275D^3)^{(0.5)}}$$

HJULSTRÖM-SUNDBORG DIAGRAM

In 1935, Filip Hjulström created the Hjulström curve, a graph which shows the relationship between the size of sediment and the velocity required to erode (lift it), transport it, or deposit it. The graph is logarithmic.

Åke Sundborg later modified the Hjulström curve to show separate curves for the movement threshold corresponding to several water depths, as is necessary if the flow velocity rather than the boundary shear stress (as in the Shields diagram) is used for the flow strength.

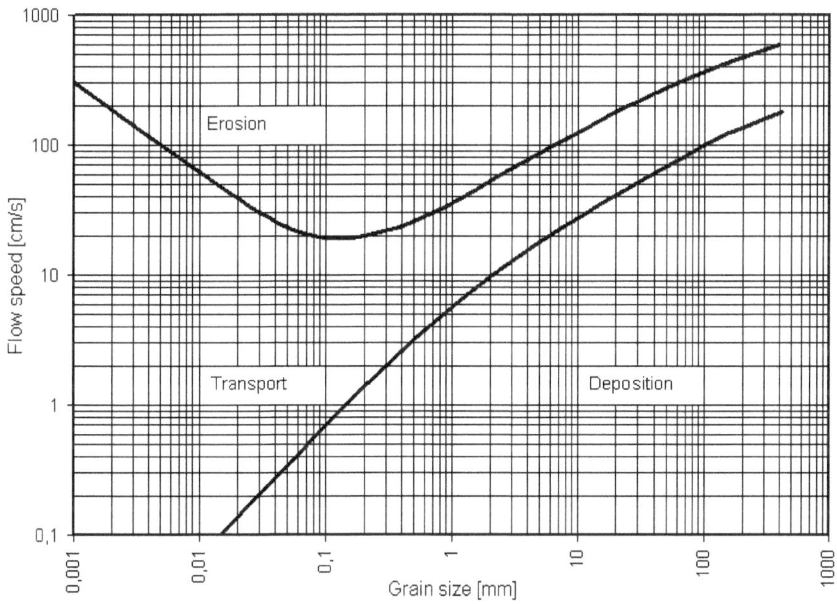

The logarithmic Hjulström curve.

This curve has no more than a historical value nowadays, although its simplicity is still attractive. Among the drawbacks of this curve are that it does not take the water depth into account and more importantly, that it does not show that sedimentation is caused by flow velocity *deceleration* and erosion is caused by flow *acceleration*. The dimensionless Shields diagram is now unanimously accepted for initiation of sediment motion in rivers. Much work was done on river sediment transport formulae in the second half of the 20th century and that work should be used preferably to Hjulström's curve, e.g. Meyer-Peter & Müller (1948), Engelund-Hansen (1967), Lefort (1991), Belleudy (2012).

Transport Rate

A schematic diagram of where the different types of sediment load are carried in the flow. Dissolved load is not sediment: it is composed of disassociated ions moving along with the flow. It may, however, constitute a significant proportion (often several percent, but occasionally greater than half) of the total amount of material being transported by the stream.

Formulas to calculate sediment transport rate exist for sediment moving in several different parts of the flow. These formulas are often segregated into bed load, suspended load, and wash load. They may sometimes also be segregated into bed material load and wash load.

Bed Load

Bed load moves by rolling, sliding, and hopping (or saltating) over the bed, and moves at a small fraction of the fluid flow velocity. Bed load is generally thought to constitute 5-10% of the total sediment load in a stream, making it less important in terms of mass balance. However, the bed material load (the bed load plus the portion of the suspended load which comprises material derived from the bed) is often dominated by bed load, especially in gravel-bed rivers. This bed material load is the only part of the sediment load that actively interacts with the bed. As the bed load is an important component of that, it plays a major role in controlling the morphology of the channel.

Bed load transport rates are usually expressed as being related to excess dimensionless shear stress raised to some power. Excess dimensionless shear stress is a nondimensional measure of bed shear stress about the threshold for motion.

$$(\tau_b^* - \tau_c^*),$$

Bed load transport rates may also be given by a ratio of bed shear stress to critical shear stress, which is equivalent in both the dimensional and nondimensional cases. This ratio is called the "transport stage" $(T_s \text{ or } \phi)$and is an important in that it shows bed shear stress as a multiple of the value of the criterion for the initiation of motion.

$$T_s = \phi = \frac{\tau_b}{\tau_c}$$

When used for sediment transport formulae, this ratio is typically raised to a power.

The majority of the published relations for bedload transport are given in dry sediment weight per unit channel width, b ("breadth"):

$$q_s = \frac{Q_s}{b}.$$

Due to the difficulty of estimating bed load transport rates, these equations are typically only suitable for the situations for which they were designed.

Notable Bed Load Transport Formulae

Meyer-Peter Müller and Derivatives

The transport formula of Meyer-Peter and Müller, originally developed in 1948, was designed for well-sorted fine gravel at a transport stage of about 8. The formula uses the above nondimensionalization for shear stress,

$$\tau * = \frac{\tau}{(\rho_s - \rho)(g)(D)},$$

and Hans Einstein's nondimensionalization for sediment volumetric discharge per unit width

$$q_s * = \frac{q_s}{D\sqrt{\frac{\rho_s - \rho}{\rho} gD}} = \frac{q_s}{Re_p \nu}.$$

Their formula reads:

$$q_s * = 8(\tau * - \tau *_c)^{3/2}.$$

Their experimentally determined value for $\tau *_c$ is 0.047, and is the third commonly used value for this (in addition to Parker's 0.03 and Shields' 0.06).

Because of its broad use, some revisions to the formula have taken place over the years that show that the coefficient on the left ("8" above) is a function of the transport stage:

$$T_s \approx 2 \rightarrow q_s * = 5.7(\tau * - 0.047)^{3/2}$$

$$T_s \approx 100 \rightarrow q_s * = 12.1(\tau * - 0.047)^{3/2}$$

The variations in the coefficient were later generalized as a function of dimensionless shear stress:

$$\begin{cases} q_s* = \alpha_s(\tau * - \tau_c*)^n \\ n = \frac{3}{2} \\ \alpha_s = 1.6\ln(\tau*) + 9.8 \approx 9.64\tau *^{0.166} \end{cases}$$

Wilcock and Crowe

In 2003, Peter Wilcock and Joanna Crowe (now Joanna Curran) published a sediment transport formula that works with multiple grain sizes across the sand and gravel range. Their formula works with surface grain size distributions, as opposed to older models which use subsurface grain size distributions (and thereby implicitly infer a surface grain sorting).

Their expression is more complicated than the basic sediment transport rules (such as that of Meyer-Peter and Müller) because it takes into account multiple grain sizes: this requires consideration of reference shear stresses for each grain size, the fraction of the total sediment supply that falls into each grain size class, and a "hiding function".

The "hiding function" takes into account the fact that, while small grains are inherently more mobile than large grains, on a mixed-grain-size bed, they may be trapped in deep pockets between large grains. Likewise, a large grain on a bed of small particles will be stuck in a much smaller pocket than if it were on a bed of grains of the same size. In gravel-bed rivers, this can cause "equal mobility", in which small grains can move just as easily as large ones. As sand is added to the system, it moves away from the "equal mobility" portion of the hiding function to one in which grain size again matters.

Their model is based on the transport stage, or ratio of bed shear stress to critical shear stress for the initiation of grain motion. Because their formula works with several grain sizes simultaneously, they define the critical shear stress for each grain size class, τ_{c,D_i}, to be equal to a "reference shear stress", τ_{ri}.

They express their equations in terms of a dimensionless transport parameter, W_i^* (with the *" indicating nondimensionality and the i" indicating that it is a function of grain size):

$$W_i^* = \frac{Rgq_{bi}}{F_i u *^3}$$

q_{bi} is the volumetric bed load transport rate of size class i per unit channel width b. F_i is the proportion of size class i that is present on the bed.

They came up with two equations, depending on the transport stage, ϕ. For $\phi < 1.35$:

$$W_i^* = 0.002\phi^{7.5}$$

and for $\phi \geq 1.35$:

$$W_i^* = 14\left(1 - \frac{0.894}{\phi^{0.5}}\right)^{4.5}.$$

This equation asymptotically reaches a constant value of W_i^* as ϕ

Wilcock and Kenworthy

In 2002, Peter Wilcock and Kenworthy T.A. , following Peter Wilcock (1998), published a sediment bed-load transport formula that works with only two sediments fractions, i.e. sand and gravel fractions. Peter Wilcock and Kenworthy T.A. in their article recognized that a mixed-sized sediment bed-load transport model using only two fractions offers practical advantages in terms of both computational and conceptual modeling by taking into account the nonlinear effects of sand presence in gravel beds on bed-load transport rate of both fractions. In fact, in the two-fraction bed load formula appears a new ingredient with respect to that of Meyer-Peter and Müller that is the proportion F_i of fraction i on the bed surface where the subscript I represents either the sand (s) or gravel (g) fraction. The proportion F_i, as a function of sand content f_s, physically represents the relative influence of the mechanisms controlling sand and gravel transport, associated with the change from a clast-supported to matrix-supported gravel bed. Moreover, since f_s spans between 0 and 1, phenomena that vary with f_s include the relative size effects producing "hiding" of fine grains and "exposure" of coarse grains. The "hiding" effect takes into account the fact that, while small grains are inherently more mobile than large grains, on a mixed-grain-size bed, they may be trapped in deep pockets between large grains. Likewise, a large grain on a bed of small particles will be stuck in a much smaller pocket than if it were on a bed of grains of the same size, which the Meyer-Peter and Müller formula refers to. In gravel-bed rivers, this can cause "equal mobility", in which small grains can move just as easily as large ones. As sand is added to the system, it moves away from the "equal mobility" portion of the hiding function to one in which grain size again matters.

Their model is based on the transport stage,*i.e.* ϕ, or ratio of bed shear stress to critical shear stress for the initiation of grain motion. Because their formula works with only two fractions simultaneously, they define the critical shear stress for each of the two grain size classes, τ_{ri}, where i represents either the sand (s) or gravel (g) fraction . The critical shear stress that represents the incipient motion for each of the two fractions is consistent with established values in the limit of pure sand and gravel beds and shows a sharp change with increasing sand content over the transition from a clast- to matrix-supported bed.

They express their equations in terms of a dimensionless transport parameter, W_i^* (with the "*" indicating nondimensionality and the "i" indicating that it is a function of grain size):

$$W_i^* = \frac{Rgq_{bi}}{F_i u*^3}$$

q_{bi} is the volumetric bed load transport rate of size class i per unit channel width b. F_i is the proportion of size class i that is present on the bed.

They came up with two equations, depending on the transport stage, ϕ. For $\phi < \phi'$:

$$W_i^* = 0.002\phi^{7.5}$$

and for $\phi \geq \phi'$:

$$W_i^* = A\left(1 - \frac{\chi}{\phi^{0.5}}\right)^{4.5}.$$

This equation asymptotically reaches a constant value of W_i^* as ϕ becomes large and the symbols A, ϕ', χ have the following values:

$A = 70, \phi' = 1.19, \chi = 0.908,$ *laboratory*

$A = 115, \phi' = 1.27, \chi = 0.923,$ *field*

In order to apply the above formulation, it is necessary to specify the characteristic grain sizes D_s for the sand portion and D_g for the gravel portion of the surface layer, the fractions F_s and F_g of sand and gravel, respectively in the surface layer, the submerged specific gravity of the sediment R and shear velocity associated with skin friction u_*

Kuhnle et al.

For the case in which sand fraction is transported by the current over and through an immobile gravel bed, Kuhnle *et al.*(2013), following the theoretical analysis done by Pellachini (2011), provides a new relationship for the bed load transport of the sand fraction when gravel particles remain at rest. It is worth mentioning that Kuhnle *et al.* (2013) applied the Wilcock and Kenworthy (2002) formula to their experimental data and found out that predicted bed load rates of sand fraction were about 10 times greater than measured and approached 1 as the sand elevation became near the top of the gravel layer. They, also, hypothesized that the mismatch between predicted and measured sand bed load rates is due to the fact that the bed shear stress used for the Wilcock and Kenworthy (2002) formula was larger than that available for transport within the gravel bed because of the sheltering effect of the gravel particles. To overcome this mismatch, following Pellachini (2011), they assumed that the variability of the bed shear stress available for the sand to be transported by the current would be some function of the so-called "Roughness Geometry Function" (RGF), which represents the gravel bed elevations distribution. Therefore, the sand bed load formula follows as:

$$q_s^* = 2.29 * 10^{-5} A(z_s)^{2.14} \left(\frac{\tau_b}{\tau_{cs}}\right)^{3.49}$$

where

$$q_s^* = \frac{q_s}{[(s-1)gD_s]^{0.5} \rho_s D_s}$$

the subscript $_s$ refers to the sand fraction, s represents the ratio ρ_s/ρ_w where ρ_s is the sand fraction density, $A(z_s)$ is the RGF as a function of the sand level z_s within the gravel bed, τ_b is the bed shear stress available for sand transport and τ_{cs} is the critical shear stress for incipient motion of the sand fraction, which was calculated graphically using the updated Shields-type relation of Miller *et al.*(1977).

Suspended Load

Suspended load is carried in the lower to middle parts of the flow, and moves at a large fraction of the mean flow velocity in the stream.

A common characterization of suspended sediment concentration in a flow is given by the Rouse Profile. This characterization works for the situation in which sediment concentration c_0 at one particular elevation above the bed z_0 can be quantified. It is given by the expression:

$$\frac{c_s}{c_0} = \left[\frac{z(h-z_0)}{z_0(h-z)}\right]^{-P/\alpha}$$

Here, z is the elevation above the bed, cs is the concentration of suspended sediment at that elevation, h is the flow depth, P is the Rouse number, and α relates the eddy viscosity for momentum K_m to the eddy diffusivity for sediment, which is approximately equal to one.

$$\alpha = \frac{K_s}{K_m} \approx 1$$

Experimental work has shown that α ranges from 0.93 to 1.10 for sands and silts.

The Rouse profile characterizes sediment concentrations because the Rouse number includes both turbulent mixing and settling under the weight of the particles. Turbulent mixing results in the net motion of particles from regions of high concentrations to low concentrations. Because particles settle downward, for all cases where the particles are not neutrally buoyant or sufficiently light that this settling velocity is negligible, there is a net negative concentration gradient as one goes upward in the flow. The Rouse Profile therefore gives the concentration profile that provides a balance between turbulent mixing (net upwards) of sediment and the downwards settling velocity of each particle.

Bed Material Load

Bed material load comprises the bed load and the portion of the suspended load that is sourced from the bed.

Three common bed material transport relations are the "Ackers-White", "Engelund-Hansen", "Yang" formulae. The first is for sand to granule-size gravel, and the second and third are for sand though Yang later expanded his formula to include fine gravel. That all of these formulae cover the sand-size range and two of them are exclusively for sand is that the sediment in sand-bed rivers is commonly moved simultaneously as bed and suspended load.

Engelund-Hansen

The bed material load formula of Engelund and Hansen is the only one to not include some kind of critical value for the initiation of sediment transport. It reads:

$$q_s* = \frac{0.05}{c_f}\tau *^{2.5}$$

where $q_s *$ is the Einstein nondimensionalization for sediment volumetric discharge per unit width, c_f is a friction factor, and $\tau *$ is the Shields stress. The Engelund-Hansen formula is one of the few sediment transport formulae in which a threshold "critical shear stress" is absent.

Wash Load

Wash load is carried within the water column as part of the flow, and therefore moves with the mean velocity of main stream. Wash load concentrations are approximately uniform in the water column. This is described by the endmember case in which the Rouse number is equal to 0 (i.e. the settling velocity is far less than the turbulent mixing velocity), which leads to a prediction of a perfectly uniform vertical concentration profile of material.

Total Load

Some authors have attempted formulations for the total sediment load carried in water. These formulas are designed largely for sand, as (depending on flow conditions) sand often can be carried as both bed load and suspended load in the same stream or shoreface.

Mass Balance

The overall balance between sediment in transport and sediment being deposited on the bed is given by the Exner equation. This expression states that the rate of increase in bed elevation due to deposition is proportional to the amount of sediment that falls out of the flow. This equation is important in that changes in the power of the flow changes the ability of the flow to carry sediment, and this is reflected in patterns of erosion and deposition observed throughout a stream. This can be localized, and simply due to small obstacles: examples are scour holes behind boulders, where flow accelerates, and deposition on the inside of meander bends. Erosion and deposition can also be regional: erosion can occur due to dam removal and base level fall. Deposition can occur due to dam emplacement that causes the river to pool, and deposit its entire load or due to base level rise.

SHORES AND SHALLOW SEAS

Seas, oceans and lakes accumulate sediment over time. The sediment could consist of *terrigenous* material, which originates on land, but may be deposited in either terrestrial, marine, or lacustrine (lake) environments; or of sediments (often biological) originating in the body of water. Terrigenous material is often supplied by nearby rivers and streams or reworked marine sediment (e.g. sand). In the mid-ocean, the exoskeletons of dead organisms are primarily responsible for sediment accumulation.

Deposited sediments are the source of sedimentary rocks, which can contain fossils of the inhabitants of the body of water that were, upon death, covered by accumulating sediment. Lake bed sediments that have not solidified into rock can be used to determine past climatic conditions.

Key Marine Depositional Environments

The major areas for deposition of sediments in the marine environment include:

- Littoral sands (e.g. beach sands, runoff river sands, coastal bars and spits, largely clastic with little faunal content)
- The continental shelf (silty clays, increasing marine faunal content).
- The shelf margin (low terrigenous supply, mostly calcareous faunal skeletons)
- The shelf slope (much more fine-grained silts and clays)
- Beds of estuaries with the resultant deposits called “bay mud”.

Holocene eolianite and a carbonate beach on Long Island, Bahamas.

One other depositional environment which is a mixture of fluvial and marine is the turbidite system, which is a major source of sediment to the deep sedimentary and abyssal basins as well as the deep oceanic trenches.

Any depression in a marine environment where sediments accumulate over time is known as a sediment trap.

The null point theory explains how sediment deposition undergoes a hydrodynamic sorting process within the marine environment leading to a seaward fining of sediment grain size.

REFERENCES

1. Anderson, R (1990). "Eolian ripples as examples of self-organization in geomorphological systems". Earth-Science Reviews. 29: 77. doi:10.1016/0012-8252(0)90029-U.
2. Kocurek, Gary; Ewing, Ryan C. (2005). "Aeolian dune field self-organization – implications for the formation of simple versus complex dune-field patterns". Geomorphology. 72: 94. Bibcode:2005Geomo..72...94K. doi:10.1016/j.geomorph.2005.05.005.
3. Goudie, A; Middleton, N.J. (2001). "Saharan dust storms: nature and consequences". Earth-Science Reviews. 56: 179. Bibcode:2001ESRv...56..179G. doi:10.1016/S0012-8252(01)00067-8.
4. http://earthobservatory.nasa.gov/IOTD/view.php?id=6458
5. Ashton, Andrew; Murray, A. Brad; Arnault, Olivier (2001). "Formation of coastline features by large-scale instabilities induced by high-angle waves". Nature. 414

(6861): 296–300. Bibcode:2001Natur.414..296A. doi:10.1038/35104541. PMID 11713526.

6. Roering, Joshua J.; Kirchner, James W.; Dietrich, William E. (1999). "Evidence for nonlinear, diffusive sediment transport on hillslopes and implications for landscape morphology". Water Resources Research. 35 (3): 853. Bibcode:1999WRR....35..853R. doi:10.1029/1998WR900090.
7. Grant, J., Walker, T.R., Hill P.S., Lintern, D.G. (2013) BEAST-A portable device for quantification of erosion in intact sediment cores. Methods in Oceanography. DOI: 10.1016/j.mio.2013.03.001
8. Shields, A. (1936) Anwendung der Ähnlichkeitsmechanik und der Turbulenzforschung auf die Geschiebebewegung; In Mitteilungen der Preussischen Versuchsanstalt für Wasserbau und Schiffbau, Heft 26 (Online; PDF; 3,8 MB)
9. Saniya Sharmeen and Garry R. Willgoose1, The interaction between armouring and particle weathering for eroding landscapes, Earth surface Processes and Landforms 31, 1195–1210 (2006)
10. Walker, T.R., Grant, J. (2009) Quantifying erosion rates and stability of bottom sediments at mussel aquaculture sites in Prince Edward Island, Canada. Journal of Marine Systems. 75: 46-55. doi:10.1016/j.jmarsys.2008.07.009
11. Dey S. (1999) Sediment threshold. Applied Mathematical Modelling, Elsevier, Vol. 23, No. 5, 399-417.
12. Hubert Chanson (2004). The Hydraulics of Open Channel Flow: An Introduction. Butterworth-Heinemann, 2nd edition, Oxford, UK, 630 pages. ISBN 978-0-7506-5978-9.
13. Whipple, Kelin (2004). Hydraulic Roughness (PDF). 12.163: Surface processes and landscape evolution. MIT OCW. Retrieved 2009-03-27.
14. Whipple, Kelin (September 2004). IV. Essentials of Sediment Transport (PDF). 12.163/12.463 Surface Processes and Landscape Evolution: Course Notes. MIT OpenCourseWare. Retrieved 2009-10-11.
15. Moore, Andrew. Lecture 20—Some Loose Ends (PDF). Lecture Notes: Fluvial Sediment Transport. Kent State. Retrieved 23 December 2009.
16. Dietrich, W. E. (1982). Settling Velocity of Natural Particles (PDF). Water Resources Research. 18 (6): 1615–1626. Bibcode:1982WRR....18.1615D. doi:10.1029/WR018i006p01615.
17. Ferguson, R. I., and M. Church (2006), A Simple Universal Equation for Grain Settling Velocity, Journal of Sedimentary Research, 74(6) 933-937, doi:10.1306/051204740933
18. The long profile – changing processes: types of erosion, transportation and deposition, types of load; the Hjulstrom curve. coolgeography.co.uk. Last accessed 26 Dec 2011.
19. Special Topics: An Introduction to Fluid Motions, Sediment Transport, and Current-generated Sedimentary Structures; As taught in: Fall 2006. Massachusetts Institute of Technology. 2006. Last accessed 26 Dec 2011.

20. Meyer-Peter, E; Müller, R. (1948). Formulas for bed-load transport. Proceedings of the 2nd Meeting of the International Association for Hydraulic Structures Research. pp. 39–64.
21. Fernandez-Luque, R; van Beek, R (1976). Erosion and transport of bedload sediment. Jour. Hyd. Research. 14 (2).
22. Cheng, Nian-Sheng (2002). Exponential Formula for Bedload Transport. Journal of Hydraulic Engineering. 128 (10): 942. doi:10.1061/(ASCE)0733-9429(2002)128:10(942).
23. Wilson, K. C. (1966). Bed-load transport at high shear stress. J. Hydraul. Div. ASCE. 92 (6): 49–59.
24. Wiberg, Patricia L.; Dungan Smith, J. (1989). Model for Calculating Bed Load Transport of Sediment. Journal of Hydraulic Engineering. 115: 101. doi:10.1061/(ASCE)0733-9429(1989)115:1(101).
25. Wilcock, Peter R.; Crowe, Joanna C. (2003). Surface-based Transport Model for Mixed-Size Sediment. Journal of Hydraulic Engineering. 129 (2): 120. doi:10.1061/(ASCE)0733-9429(2003)129:2(120).
26. Parker, G.; Klingeman, P. C.; McLean, D. G. (1982). Bedload and Size Distribution in Paved Gravel-Bed Streams. Journal of the Hydraulics Division. ASCE. 108 (4): 544–571.
27. Wilcock, P. R. (1998). Two-fraction model of initial sediment motion in gravel-bed rivers. Science (280): 410–412. Bibcode:1998Sci...280..410W. doi:10.1126/science.280.5362.410.
28. Wilcock, Peter R.; Kenworthy, T. (2002). A two-fraction model for the transport of sand/gravel mixtures. Water Resour. Res. 38 (10): 1194. Bibcode:2002WRR....38.1194W. doi:10.1029/2001WR000684.
29. Kuhnle, R. A.; Wren, D. G.; Langendoen, E. J.; Rigby, J. R. (2013). Sand Transport over an Immobile Gravel Substrate. Journal of Hydraulic Engineering. 139 (2). doi:10.1061/(ASCE)HY.1943-7900.0000615.
30. Pellachini, Corrado (2011). Modelling fine sediment transport over an immobile gravel bed. Trento: Unitn-eprints.
31. Nikora, V; Goring, D; McEwan, I; Griffiths, G (2001). Spatially averaged open-channel flow over rough bed. J. Hydraul. Eng. 127 (2). doi:10.1061/(ASCE)0733-9429(2001)127:2(123).
32. Miller, M.C.; McCave, I.N.; Komar, P.D. (1977). Threshold of sediment motion under unidirectional currents.. Sedimentology. 24 (4): 507–527. Bibcode:1977Sedim..24..507M. doi:10.1111/j.1365-3091.1977.tb00136.x.
33. Harris, Courtney K. (March 18, 2003). Lecture 9: Suspended Sediment Transport II (PDF). Sediment transport processes in coastal environments. Virginia Institute of Marine Science. Retrieved 23 December 2009.
34. Moore, Andrew. Lecture 21—Suspended Sediment Transport (PDF). Lecture Notes: Fluvial Sediment Transport. Kent State. Retrieved 25 December 2009.
35. Ackers, P.; White, W.R. (1973). Sediment Transport: New Approach and Analysis.

Journal of the Hydraulics Division. ASCE. 99 (11): 2041–2060.

36. Ariffin, J.; A.A. Ghani; N.A. Zakaira; A.H. Yahya (14–16 October 2002). Evaluation of equations on total bed material load (PDF). International Conference on Urban Hydrology for the 21st Century. Kuala Lumpur.
37. Yang, C (1979). Unit stream power equations for total load. Journal of Hydrology. 40: 123. Bibcode:1979JHyd...40..123Y. doi:10.1016/0022-1694(79)90092-1.
38. Bailard, James A. (1981). An Energetics Total Load Sediment Transport Model For a Plane Sloping Beach. Journal of Geophysical Research. 86: 10938. Bibcode:1981JGR....8610938B. doi:10.1029/JC086iC11p10938.

CHAPTER 5

Fluvial Deposits

Chapter Contents

- **Introduction**
- **Fluvial Processes**
- **Fluvial Terrace**
- **Lacustrine Deposits**

INTRODUCTION

Fluvial is a term used in geography and geology to refer to the processes associated with rivers and streams and the deposits and landforms created by them. When the stream or rivers are associated with glaciers, ice sheets, or ice caps, the term glaciofluvial or fluvioglacial is used.

Deep, eroding glaciofluvial deposits alongside the Matanuska River, Alaska.

FLUVIAL PROCESSES

Fluvial processes include the motion of sediment and erosion or deposition on the river bed.

Erosion by moving water can happen in two ways. Firstly, the movement of water across the bed exerts a shear stress directly onto the bed. If the cohesive strength of the substrate is lower than the shear exerted, or the bed is composed of loose sediment which can be mobilized by such stresses, then the bed will be lowered purely by clearwater flow. However, if the river carries significant quantities of sediment, this material can act as tools to enhance wear of the bed (abrasion). At the same time the fragments themselves are ground down, becoming smaller and more rounded (attrition).

Sediment in rivers is transported as either bedload (the coarser fragments which move close to the bed) or suspended load (finer fragments carried in the water). There is also a component carried as dissolved material.

For each grain size there is a specific velocity at which the grains start to move, called *entrainment velocity*. However the grains will continue to be transported even if the velocity falls below the entrainment velocity due to the reduced (or removed) friction between the grains and the river bed. Eventually the velocity will fall low enough for the grains to be deposited. This is shown by the Hjulstrøm curve.

A river is continually picking up and dropping solid particles of rock and soil from its bed throughout its length. Where the river flow is fast, more particles are picked up than dropped. Where the river flow is slow, more particles are dropped than picked up. Areas where more particles are dropped are called alluvial or flood plains, and the dropped particles are called alluvium.

Even small streams make alluvial deposits, but it is in the flood plains and deltas of large rivers that large, geologically-significant alluvial deposits are found.

The amount of matter carried by a large river is enormous. The names of many rivers derive from the color that the transported matter gives the water. For example, the Huang He in China is literally translated “Yellow River”, and the Mississippi River in the United States is also called “the Big Muddy.” It has been estimated that the Mississippi River annually carries 406 million tons of sediment to the sea, the Yellow River 796 million tons, and the Po River in Italy 67 million tons.

Bradshaw Model

The Bradshaw Model is a geographical model which describes how a river’s characteristics vary between the *upper course* and *lower course* of a river. It shows that discharge, occupied channel width, channel depth and average load quantity increase downstream. Load particle size, channel bed roughness and gradient are all characteristics that decrease. This is represented by triangles; an increase in the size of a triangle represents an increase in the variable. Generally it shows the characteristics we expect to see in a river, but due to the nature of rivers and the ever changing environment in which we live not all rivers fit the model perfectly; therefore the model is usually used in order to compare natural rivers to concepts laid down by the model.

Erosion

In earth science, erosion is the action of surface processes (such as water flow or wind) that remove soil, rock, or dissolved material from one location on the Earth's crust, then transport it away to another location. The particulate breakdown of rock or soil into clastic sediment is referred to as *physical* or *mechanical* erosion; this contrasts with *chemical* erosion, where soil or rock material is removed from an area by its dissolving into a solvent (typically water), followed by the flow away of that solution. Eroded sediment or solutes may be transported just a few millimetres, or for thousands of kilometres.

Natural rates of erosion are controlled by the action of geomorphic drivers, such as rainfall; bedrock wear in rivers; coastal erosion by the sea and waves; glacial plucking, abrasion, and scour; areal flooding; wind abrasion; groundwater processes; and mass movement processes in steep landscapes like landslides and debris flows. The rates at which such processes act control how fast a surface is eroded. Typically, physical erosion proceeds fastest on steeply sloping surfaces, and rates may also be sensitive to some climatically-controlled properties including amounts of water supplied (e.g., by rain), storminess, wind speed, wave fetch, or atmospheric temperature (especially for some ice-related processes). Feedbacks are also possible between rates of erosion and the amount of eroded material that is already carried by, for example, a river or glacier. Processes of erosion that produce sediment or solutes from a place contrast with those of deposition, which control the arrival and emplacement of material at a new location.

While erosion is a natural process, human activities have increased by 10-40 times the rate at which erosion is occurring globally. Excessive (or accelerated) erosion causes both "on-site" and "off-site" problems. On-site impacts include decreases in agricultural productivity and (on natural landscapes) ecological collapse, both because of loss of the nutrient-rich upper soil layers. In some cases, the eventual end result is desertification. Off-site effects include sedimentation of waterways and eutrophication of water bodies, as well as sediment-related damage to roads and houses. Water and wind erosion are the two primary causes of land degradation; combined, they are responsible for about 84% of the global extent of degraded land, making excessive erosion one of the most significant environmental problems worldwide.

Intensive agriculture, deforestation, roads, anthropogenic climate change and urban sprawl are amongst the most significant human activities in regard to their effect on stimulating erosion.However, there are many prevention and remediation practices that can curtail or limit erosion of vulnerable soils.

An actively eroding rill on an intensively-farmed field in eastern Germany.

A natural arch produced by the wind erosion of differentially weathered rock in Jebel Kharaz, Jordan.

A wave-like sea cliff produced by coastal erosion, in Jinshitan Coastal National Geopark, Dalian, Liaoning Province, China.

Physical Processes

Rainfall and Surface Runoff

Rainfall, and the surface runoff which may result from rainfall, produces four main types of soil erosion: *splash erosion*, *sheet erosion*, *rill erosion*, and *gully erosion*. Splash erosion is generally seen as the first and least severe stage in the soil erosion process, which is followed by sheet erosion, then rill erosion and finally gully erosion (the most severe of the four).

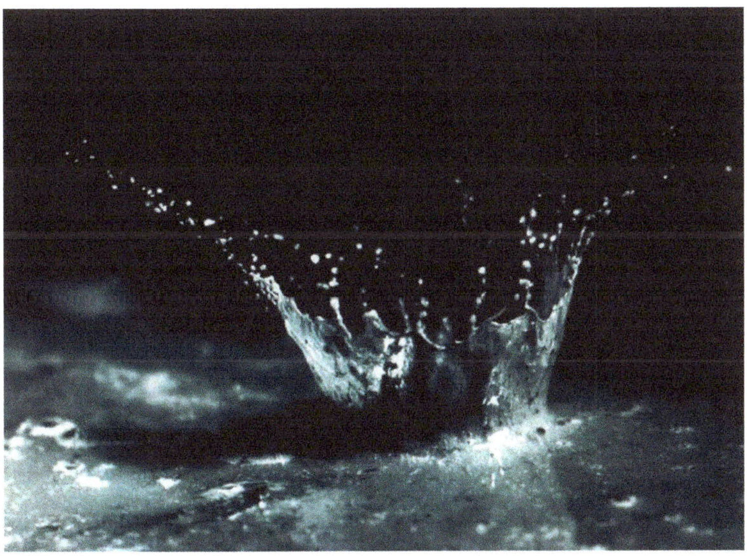

Soil and water being splashed by the impact of a single raindrop.

In *splash erosion*, the impact of a falling raindrop creates a small crater in the soil, ejecting soil particles. The distance these soil particles travel can be as much as 0.6 m (two feet) vertically and 1.5 m (five feet) horizontally on level ground.

If the soil is saturated, or if the rainfall rate is greater than the rate at which water can infiltrate into the soil, surface runoff occurs. If the runoff has sufficient flow energy, it will transport loosened soil particles (sediment) down the slope. *Sheet erosion* is the transport of loosened soil particles by overland flow.

A spoil tip covered in rills and gullies due to erosion processes caused by rainfall: Rummu, Estonia.

Rill erosion refers to the development of small, ephemeral concentrated flow paths which function as both sediment source and sediment delivery systems for erosion on hillslopes. Generally, where water erosion rates on disturbed upland areas are greatest, rills are active. Flow depths in rills are typically of the order of a few centimeters (about an inch) or less and along-channel slopes may be quite steep. This means that rills exhibit hydraulic physics very different from water flowing through the deeper, wider channels of streams and rivers.

Gully erosion occurs when runoff water accumulates and rapidly flows in narrow channels during or immediately after heavy rains or melting snow, removing soil to a considerable depth.

Rivers and Streams

Valley or *stream erosion* occurs with continued water flow along a linear feature. The erosion is both downward, deepening the valley, and headward, extending the valley into the hillside, creating head cuts and steep banks. In the earliest stage of stream erosion, the erosive activity is dominantly vertical, the valleys have a typical **V** cross-section and the stream gradient is relatively steep. When some base level is reached, the erosive activity switches to lateral erosion, which widens the valley floor and creates a narrow floodplain. The stream gradient becomes nearly flat, and lateral deposition of sediments becomes important as the stream meanders across the valley floor. In all stages of stream erosion, by far the most erosion occurs during times of flood, when more and faster-moving water is available to carry a larger sediment load. In such processes, it is not the water alone that erodes: suspended abrasive particles, pebbles and boulders can also act erosively as they traverse a surface, in a process known as *traction*.

Dobbingstone Burn, Scotland, showing two different types of erosion affecting the same place. Valley erosion is occurring due to the flow of the stream, and the boulders and stones (and much of the soil) that are lying on the stream's banks are glacial till that was left behind as ice age glaciers flowed over the terrain.

Bank erosion is the wearing away of the banks of a stream or river. This is distinguished from changes on the bed of the watercourse, which is referred to as *scour*. Erosion and changes in the form of river banks may be measured by inserting metal rods into the bank and marking the position of the bank surface along the rods at different times.

Thermal erosion is the result of melting and weakening permafrost due to moving water. It can occur both along rivers and at the coast. Rapid river channel migration observed in the Lena River of Siberia is due to thermal erosion, as these portions of the banks are composed of permafrost-cemented non-cohesive materials. Much of this erosion occurs as the weakened banks fail in large slumps. Thermal erosion also affects the Arctic coast, where wave action and near-shore temperatures combine to undercut permafrost bluffs along the shoreline and cause them to fail. Annual erosion rates along a 100-kilometre (62-mile) segment of the Beaufort Sea shoreline averaged 5.6 meters (18 feet) per year from 1955 to 2002.

Coastal Erosion

Wave cut platform caused by erosion of cliffs by the sea, at Southerndown in South Wales.

Erosion of the boulder clay (of Pleistocene age) along cliffs of Filey Bay, Yorkshire, England.

Shoreline erosion, which occurs on both exposed and sheltered coasts, primarily occurs through the action of currents and waves but sea level (tidal) change can also play a role.

Hydraulic action takes place when air in a joint is suddenly compressed by a wave closing the entrance of the joint. This then cracks it. *Wave pounding* is when the sheer energy of the wave hitting the cliff or rock breaks pieces off. *Abrasion* or *corrasion* is caused by waves launching seaload at the cliff. It is the most effective and rapid form of shoreline erosion (not to be confused with *corrosion*). *Corrosion* is the dissolving of rock by carbonic acid in sea water. Limestone cliffs are particularly vulnerable to this kind of erosion. *Attrition* is where particles/seaload carried by the waves are worn down as they hit each other and the cliffs. This then makes the material easier to wash away. The material ends up as shingle and sand. Another significant source of erosion, particularly on carbonate coastlines, is the boring, scraping and grinding of organisms, a process termed *bioerosion*.

Sediment is transported along the coast in the direction of the prevailing current (longshore drift). When the upcurrent amount of sediment is less than the amount being carried away, erosion occurs. When the upcurrent amount of sediment is greater, sand or gravel banks will tend to form as a result of deposition. These banks may slowly migrate along the coast in the direction of the longshore drift, alternately protecting and exposing parts of the coastline. Where there is a bend in the coastline, quite often a buildup of eroded material occurs forming a long narrow bank (a spit). Armoured beaches and submerged offshore sandbanks may also protect parts of a coastline from erosion. Over the years, as the shoals gradually shift, the erosion may be redirected to attack different parts of the shore.

Chemical Erosion

Chemical erosion is the loss of matter in a landscape in the form of solutes. Chemical erosion is usually calculated from the solutes found in streams. Anders Rapp pioneered the study of chemical erosion in his work about Kärkevagge published in 1960.

Glaciers

Glaciers erode predominantly by three different processes: abrasion/scouring, plucking, and ice thrusting. In an abrasion process, debris in the basal ice scrapes along the bed, polishing and gouging the underlying rocks, similar to sandpaper on wood. Glaciers can also cause pieces of bedrock to crack off in the process of plucking. In ice thrusting, the glacier freezes to its bed, then as it surges forward, it moves large sheets of frozen sediment at the base along with the glacier. This method produced some of the many thousands of lake basins that dot the edge of the Canadian Shield. The erosion caused by glaciers worldwide erodes mountains so effectively that the term *glacial buzz-saw* has become widely used, which describes the limiting effect of glaciers on the height of mountain ranges. As mountains grow higher, they generally allow for more glacial activity (especially in the accumulation zone above the glacial equilibrium line altitude),which causes increased rates of erosion of the mountain, decreasing mass faster than isostatic rebound can add to the mountain. This provides a good example of a negative feedback loop. Ongoing research is showing that while glaciers tend to decrease mountain size, in some areas, glaciers can actually reduce the rate of erosion, acting as a *glacial armour*.

Glacial moraines above Lake Louise, in Alberta, Canada.

These processes, combined with erosion and transport by the water network beneath the glacier, leave moraines, drumlins, ground moraine (till), kames, kame deltas, moulins, and glacial erratics in their wake, typically at the terminus or during glacier retreat.

Floods

At extremely high flows, kolks, or vortices are formed by large volumes of rapidly rushing water. Kolks cause extreme local erosion, plucking bedrock and creating pothole-type geographical features called Rock-cut basins. Examples can be seen in the flood regions result from glacial Lake Missoula, which created the channeled scablands in the Columbia Basin region of eastern Washington.

Wind Erosion

Wind erosion is a major geomorphological force, especially in arid and semi-arid regions. It is also a major source of land degradation, evaporation, desertification, harmful airborne dust, and crop damage—especially after being increased far above natural rates by human activities such as deforestation, urbanization, and agriculture.

Wind erosion is of two primary varieties: *deflation*, where the wind picks up and carries away loose particles; and *abrasion*, where surfaces are worn down as they are struck by airborne particles carried by wind. Deflation is divided into three categories: (1) *surface creep*, where larger, heavier particles slide or roll along the ground; (2) *saltation*, where particles are lifted a short height into the air, and bounce and saltate across the surface of the soil; and (3) *suspension*, where very small and light particles are lifted into the air by the wind, and are often carried for long distances. Saltation is responsible for the majority (50-70%) of wind erosion, followed by suspension (30-40%), and then surface creep (5-25%).

Árbol de Piedra, a rock formation in the Altiplano, Bolivia sculpted by wind erosion.

Wind erosion is much more severe in arid areas and during times of drought. For example, in the Great Plains, it is estimated that soil loss due to wind erosion can be as much as 6100 times greater in drought years than in wet years.

Mass Movement

Wadi in Makhtesh Ramon, Israel, showing gravity collapse erosion on its banks.

Mass movement is the downward and outward movement of rock and sediments on a sloped surface, mainly due to the force of gravity.

Mass movement is an important part of the erosional process, and is often the first stage in the breakdown and transport of weathered materials in mountainous areas. It moves material from higher elevations to lower elevations where other eroding agents such as streams and glaciers can then pick up the material and move it to even lower elevations. Mass-movement processes are always occurring continuously on all slopes; some mass-movement processes act very slowly; others occur very suddenly, often with disastrous results. Any perceptible down-slope movement of rock or sediment is often referred to in general terms as a landslide. However, landslides can be classified in a much more detailed way that reflects the mechanisms responsible for the movement and the velocity at which the movement occurs. One of the visible topographical manifestations of a very slow form of such activity is a scree slope.

Slumping happens on steep hillsides, occurring along distinct fracture zones, often within materials like clay that, once released, may move quite rapidly downhill. They will often show a spoon-shaped isostatic depression, in which the material has begun to slide downhill. In some cases, the slump is caused by water beneath the slope weakening it. In many cases it is simply the result of poor engineering along highways where it is a regular occurrence.

Surface creep is the slow movement of soil and rock debris by gravity which is usually not perceptible except through extended observation. However, the term can also describe the rolling of dislodged soil particles 0.5 to 1.0 mm (0.02 to 0.04 in) in diameter by wind along the soil surface.

Factors affecting erosion rates

Climate

The amount and intensity of precipitation is the main climatic factor governing soil erosion by water. The relationship is particularly strong if heavy rainfall occurs at times when, or in locations where, the soil's surface is not well protected by vegetation. This might be during periods when agricultural activities leave the soil bare, or in semi-arid regions where vegetation is naturally sparse. Wind erosion requires strong winds, particularly during times of drought when vegetation is sparse and soil is dry (and so is more erodible). Other climatic factors such as average temperature and temperature range may also affect erosion, via their effects on vegetation and soil properties. In general, given similar vegetation and ecosystems, areas with more precipitation (especially high-intensity rainfall), more wind, or more storms are expected to have more erosion.

In some areas of the world (e.g. the mid-western USA), rainfall intensity is the primary determinant of erosivity, with higher intensity rainfall generally resulting in more soil erosion by water. The size and velocity of rain drops is also an important factor. Larger and higher-velocity rain drops have greater kinetic energy, and thus their impact will displace soil particles by larger distances than smaller, slower-moving rain drops.

In other regions of the world (e.g. western Europe), runoff and erosion result from relatively low intensities of stratiform rainfall falling onto previously saturated soil. In such situations, rainfall amount rather than intensity is the main factor determining the severity of soil erosion by water.

Vegetative Cover

Vegetation acts as an interface between the atmosphere and the soil. It increases the permeability of the soil to rainwater, thus decreasing runoff. It shelters the soil from winds, which results in decreased wind erosion, as well as advantageous changes in microclimate. The roots of the plants bind the soil together, and interweave with other roots, forming a more solid mass that is less susceptible to both water and wind erosion. The removal of vegetation increases the rate of surface erosion.

Topography

The topography of the land determines the velocity at which surface runoff will flow, which in turn determines the erosivity of the runoff. Longer, steeper slopes (especially those without adequate vegetative cover) are more susceptible to very high rates of erosion during heavy rains than shorter, less steep slopes. Steeper terrain is also more prone to mudslides, landslides, and other forms of gravitational erosion processes.

Tectonics

Tectonic processes control rates and distributions of erosion at the Earth's surface. If tectonic action causes part of the Earth's surface (e.g., a mountain range) to be raised or lowered relative to surrounding areas, this must necessarily change the gradient of the land surface. Because erosion rates are almost always sensitive to local slope (see above), this will change the rates of erosion in the uplifted area. Active tectonics also brings fresh, unweathered rock towards the surface, where it is exposed to the action of erosion.

However, erosion can also affect tectonic processes. The removal by erosion of large amounts of rock from a particular region, and its deposition elsewhere, can result in a lightening of the load on the lower crust and mantle. Because tectonic processes are driven by gradients in the stress field developed in the crust, this unloading can in turn cause tectonic or isostatic uplift in the region. In some cases, it has been hypothesised that these twin feedbacks can act to localise and enhance zones of very rapid exhumation of deep crustal rocks beneath places on the Earth's surface with extremely high erosion rates, for example, beneath the extremely steep terrain of Nanga Parbat in the western Himalayas. Such a place has been called a "tectonic aneurysm".

Erosion of Earth systems

Mountain Ranges

Mountain ranges are known to take many millions of years to erode to the degree they effectively cease to exist. Scholars Pitman and Golovchenko estimate that it takes probably more than 450 million years to erode a mountain mass similar to the Himalaya into an almost-flat peneplain if there are no major sea-level changes. Erosion of mountains massifs can create a pattern of equally high summits called summit accordance. It has been argued that extension during post-orogenic collapse is a more effective mechanism of lowering the height of orogenic mountains than erosion.

Examples of heavily eroded mountain ranges include the Timanides of Northern Russia. Erosion of this orogen has produced sediments that are now found in the East European Platform, including the Cambrian Sablya Formation near Lake Ladoga. Studies of these sediments points that its likely that the erosion of the orogen was beginning in the Cambrian and then became stronger in Ordovician.

Soils

If the rate of erosion is higher than the rate of soil formation the soils are being destroyed by erosion. Where soil is not destroyed by erosion, erosion can in some cases prevent the formation of soil features that form slowly. Inceptisols are common soils that form in areas of fast erosion.

While erosion of soils is a natural process, human activities have increased by 10-40 times the rate at which erosion is occurring globally. Excessive (or accelerated) erosion causes both "on-site" and "off-site" problems. On-site impacts include decreases in agricultural productivity and (on natural landscapes) ecological collapse, both because of loss of the nutrient-rich upper soil layers. In some cases, the eventual end result is desertification. Off-site effects include sedimentation of waterways and eutrophication of water bodies, as well as sediment-related damage to roads and houses. Water and wind erosion are the two primary causes of land degradation; combined, they are responsible for about 84% of the global extent of degraded land, making excessive erosion one of the most significant environmental problems worldwide.

Suspension

A suspension is a heterogeneous mixture containing solid particles that are sufficiently large for sedimentation. Usually they must be larger than one micrometer. A Suspension is a heterogeneous mixture in which the solute particles do not dissolve but get suspended throughout the bulk of the medium. Particles of suspension are visible to the naked eye. It is when particles are left floating around freely in a solvent. The internal phase (solid) is dispersed throughout the external phase (fluid) through mechanical agitation, with the use of certain excipients or suspending agents. Unlike colloids, suspensions will eventually settle. An example of a suspension would be sand in water. The suspended particles are visible under a microscope and will settle over time if left undisturbed. This distinguishes a suspension from a colloid, in which the suspended particles are smaller and do not settle. Colloids and suspensions are different from solutions, in which the dissolved substance (solute) does not exist as a solid, and solvent and solute are homogeneously mixed.

A suspension of liquid droplets or fine solid particles in a gas is called an aerosol or particulate. In the atmosphere these consist of fine dust and soot particles, sea salt, biogenic and volcanogenic sulfates, nitrates, and cloud droplets.

Suspensions are classified on the basis of the dispersed phase and the dispersion medium, where the former is essentially solid while the latter may either be a solid, a liquid, or a gas.

In modern chemical process industries, high shear mixing technology has been used to create many novel suspensions.

Suspensions are unstable from the thermodynamic point of view; however, they can be kinetically stable over a large period of time, which determines their shelf life. This time span needs to be measured to ensure the best product quality to the final consumer. "Dispersion stability refers to the ability of a dispersion to resist change in its properties over time." D.J. McClements.

Technique Monitoring Physical Stability

Multiple light scattering coupled with vertical scanning is the most widely used technique to monitor the dispersion state of a product, hence identifying and quantifying destabilization phenomena. It works on concentrated dispersions without dilution. When light is sent through the sample, it is back scattered by the particles. The backscattering intensity is directly proportional to the size and volume fraction of the dispersed phase. Therefore, local changes in concentration (sedimentation) and global changes in size (flocculation, aggregation) are detected and monitored.

Accelerating Methods for Shelf Life Prediction

The kinetic process of destabilization can be rather long (up to several months or even years for some products) and it is often required for the formulator to use further accelerating methods in order to reach reasonable development time for new product design. Thermal methods are the most commonly used and consists in increasing temperature to accelerate destabilization (below critical temperatures of phase inversion or chemical degradation). Temperature affects not only the viscosity, but also interfacial tension in the case of non-ionic surfactants or more generally interactions forces inside the system. Storing a dispersion at high temperatures enables simulation of real life conditions for a product (e.g. tube of sunscreen cream in a car in the summer), but also to accelerate destabilization processes up to 200 times.

Mechanical acceleration including vibration, centrifugation and agitation are sometimes used. They subject the product to different forces that pushes the particles / droplets against one another, hence helping in the film drainage. However, some emulsions would never coalesce in normal gravity, while they do under artificial gravity. Moreover, segregation of different populations of particles have been highlighted when using centrifugation and vibration.

Common examples of suspensions include:

- Mud or muddy water: where soil, clay, or silt particles are suspended in water
- Flour suspended in water

FLUVIAL TERRACE

Fluvial terraces are elongated terraces that flank the sides of floodplains and fluvial valleys all over the world. They consist of a relatively level strip of land, called a "tread," separated from either an adjacent floodplain, other fluvial terraces, or uplands by distinctly steeper strips of land called "risers." These terraces lie parallel to and above the river channel and its floodplain. Because of the manner in which they form, fluvial terraces are underlain by fluvial sediments of highly variable thickness.

Fluvial terraces are the remnants of earlier floodplains that existed at a time when either a stream or river was flowing at a higher elevation before its channel downcut to create a new floodplain at a lower elevation. Changes in elevation can be due to changes in the base level (elevation of the lowest point in the fluvial system, usually the drainage basin) of the fluvial system, which leads to headward erosion along the length of either a stream or river, gradually lowering its elevation. For example, downcutting by a river can lead to increased velocity of a tributary, causing that tributary to erode toward its headwaters. Terraces can also be left behind when the volume of the fluvial flow declines due to changes in climate, typical of areas which were covered by ice during periods of glaciation, and their adjacent drainage basins.

Types

There are two basic types of fluvial terraces, *fill terraces* and *strath terraces.* Fill terraces sometimes are further subdivided into *nested fill terraces* and *cut terraces.* Both fill and strath terraces are, at times, described as being either *paired* or *unpaired* terraces based upon the relative elevations of the surface of these terraces.

Eroded alluvial fill 60 feet thick at Kanab Creek, Kane County, Utah. In 1884 the stream ran at top of the terrace. 1939 photo by United States Geological Survey.

Fill terraces: Fill terraces are the result of an existing valley being filled with alluvium. The valley may fill with alluvium for many different reasons including: an influx in bed load due to glaciation or change in stream power which causes the valley, that was down cut by either a stream or river, to be filled in with material (Easterbrook).

The stream or river will continue to deposit material until an equilibrium is reached and it can transport the material rather than deposit it. This equilibrium may last for a very short period, such as, after glaciation, or for a very long time if the conditions do not change. The fill terrace is created when the conditions change again and either a stream or river starts to incise into the material that it deposited in the valley. Once this occurs benches composed completely of alluvium form on the sides of the valley. The upper most benches are the fill terraces. As it continues to cut down through the alluvium the fill terraces are left above the river channel (sometimes 100 m or more). The fill terrace is only the very highest terrace resulting from the depositional episode, if there are multiple terraces below the fill terrace these are called cut terraces.

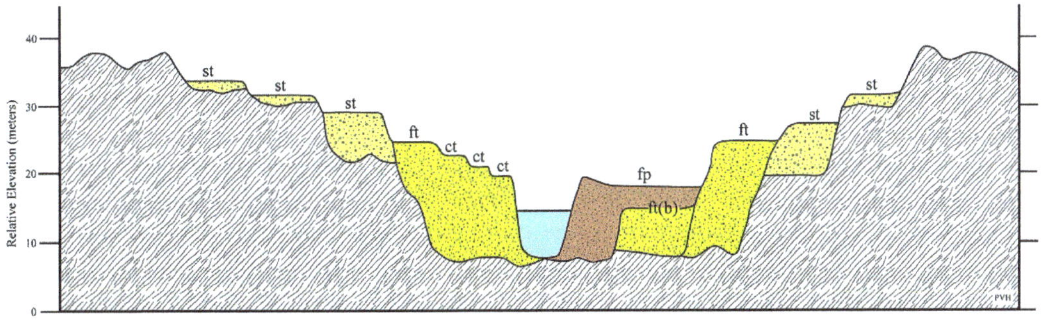

Hypothetical valley cross-section illustrating a complex sequence of aggradational (fill) and degradational (cut and strath) terraces. Note ct = cut terrace, ft = fill terrace, ft(b) = buried fill terrace, fp = active floodplain, and st = strath terrace.

Cut terraces: Cut terraces, also called "cut-in-fill" terraces, are similar to the fill terraces mentioned above, but they are erosional in origin. Once the alluvium deposited in the valley has begun to erode and fill terraces form along the valley walls, cut terraces may also form below the fill terraces. As either a stream or river continues to incise into the material, multiple levels of terraces may form. The uppermost being the fill terraces and the remaining lower terraces are cut terraces.

Nested fill terraces: Nested fill terraces are the result of the valley filling with alluvium, the alluvium being incised, and the valley filling again with material but to a lower level than before. The terrace that results for the second filling is a nested terrace because it has been "nested" into the original alluvium and created a terrace. These terraces are depositional in origin and may be able to be identified by a sudden change in alluvium characteristics such as finer material for example.

Strath terraces: Strath terraces are the result of either a stream or river downcutting through bedrock. As the flow continues to downcut, a period of valley widening may occur and expand the valley width. This may occur due to an equilibrium reached in the fluvial system resulting from: slowed or paused uplift, climate change, or a change in the bedrock type. Once downcutting continues the flattened valley bottom composed of bedrock (overlain with a possible thin layer of alluvium) is left above either a stream or river channel. These bedrock terraces are the strath terraces and are erosional in nature.

Unpaired fluvial terraces on the South Fork of the Shoshone River, Park County, Wyoming, 1923. The river at left has encountered a formation of erosion-resistant volcanic breccia, causing it to downcut more rapidly on the right, leaving terraces of different elevations.

Paired and unpaired terraces: Terraces of the same elevation on opposite sides of either a stream or river are called *paired terraces*. They occur when it downcuts evenly on both sides and terraces on one side of the river correspond in height with those on the other side. Paired terraces are caused by river rejuvenation. *Unpaired terraces* occur when either a stream or river encounters material on one side that resists erosion, leaving a single terrace with no corresponding terrace on the resistant side.

Applications

Fluvial terraces can be used to measure the rate at which either a stream or river is downcutting its valley. Using various dating methods, an age can be determined for the deposition of the terrace. Using the resulting date and the elevation above its current level, an approximate average rate of downcutting can be determined.

LACUSTRINE DEPOSITS

Lacustrine deposits are sedimentary rock formations which formed in the bottom of ancient lakes. A common characteristic of lacustrine deposits is that a river or stream channel has carried sediment into the basin. Lacustrine deposits form in all lake types including rift graben lakes, oxbow lakes, glacial lakes, and crater lakes. Lacustrine environments, like seas, are large bodies of water. They share similar sedimentary deposits which are mainly composed of low-energy particle sizes. Lacustrine deposits are typically very well sorted with highly laminated beds of silts, clays, and occasionally carbonates. In regards to geologic time, lakes are temporary and once they no longer receive water, they dry up and leave a formation.

Lake Types

Lacustrine deposits can form in every variety of basins found in nature. How each basin originates is where the distinction between lacustrine deposit types stem. Rift graben lakes are formed from crustal stretching also known as rifting. Sediment influx is typically dominated by precipitation runoff and discharge through channels migrating towards the depression. Oxbow lakes form lacustrine deposits from seasonal overbank flooding as well as precipitation runoff which refills these isolated basins with fresh water and new sediments. Glacial lakes form when terminal moraines block water from escaping the newly carved valley from glacial erosion. As the glacier melts, the valley fills with melt water that creates a glacial lake. Crater lakes can be meteoritic or of the caldera variety. Crater lakes sediments are provided from precipitation runoff descending their steep slopes.

Resource Value

Lacustrine deposits have gained more attention recently due to containing valuable source rocks of oil, coal, and uranium. Lacustrine deposits generally provide productive mining conditions but can prove challenging when underground mines are attempted due to the poor shear strength of clays and silts as well as the amount of moisture often locked in the layers due to a low permeability characteristic of lacustrine deposits. Ephemeral lakes have recently been found to be especially valuable due to their seasonal wetting and drying out characteristic. Between the wet and the dry seasons, there is a time period when organic matter has the perfect opportunity to generate, only to be submerged by the influx of water during the wet season. This action creates a prime combination of organic matter, silts and clays to create oil shales or coal deposits.

Fluvial Processes: Rivers, Streams, and Overland Flow

Particle Motion

Rivers and streams carry sediment in their flows. This sediment can be in a variety of locations within the flow, depending on the balance between the upwards velocity on the particle (drag and lift forces), and the settling velocity of the particle. These relationships are shown in the following table for the Rouse number, which is a ratio of sediment fall velocity to upwards velocity.

$$\textbf{Rouse} = \frac{\text{Settling velocity}}{\text{Upwards velocity from lift and drag}} = \frac{w_s}{\kappa u_*}$$

where

- w_s is the fall velocity
- κ is the von Kármán constant
- u_* is the shear velocity

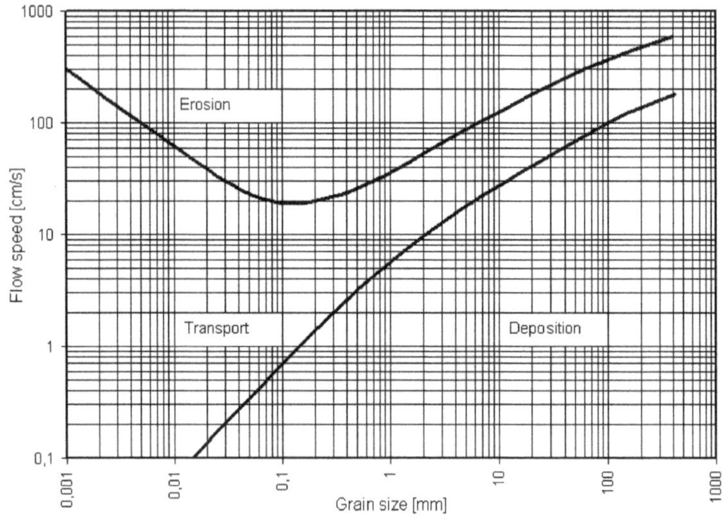

Hjulström curve: The velocities of currents required for erosion, transportation, and deposition (sedimentation) of sediment particles of different sizes.

Mode of Transport	Rouse Number
Bed load	>2.5
Suspended load: 50% Suspended	>1.2, <2.5
Suspended load: 100% Suspended	>0.8, <1.2
Wash load	<0.8

If the upwards velocity approximately equal to the settling velocity, sediment will be transported downstream entirely as suspended load. If the upwards velocity is much less than the settling velocity, but still high enough for the sediment to move, it will move along the bed as bed load by rolling, sliding, and saltating (jumping up into the flow, being transported a short distance then settling again). If the upwards velocity is higher than the settling velocity, the sediment will be transported high in the flow as wash load.

As there are generally a range of different particle sizes in the flow, it is common for material of different sizes to move through all areas of the flow for given stream conditions.

Fluvial Bedforms

A bedform is a feature that develops at the interface of fluid and a moveable bed, the result of bed material being moved by fluid flow. Examples include ripples and dunes on the bed of a river. Bedforms are often preserved in the rock record as a result of being present in a depositional setting. Bedforms are often characteristic to the flow parameters, and may be used to infer flow depth and velocity, and therefore the Froude number.

Current ripples preserved in sandstone of the Moenkopi Formation, Capitol Reef National Park, Utah, United States.

Bedforms Initiation

Bedforms are omnipresent in many environments (e.g., fluvial, eolian, glaciofluvial, deltaic and deep sea), although there is still some debate on how they develop. The initiation of bedforms have two - not mutually exclusive - models: defect initiation and instantaneous initiation.

Defect Initiation

The defect theory proposes that the turbulent sweeps that are generated in turbulent flows entrain sediment that upon deposition generates defects in a non-cohesive material. These deposits then propagate downstream via a flow separation process, thus developing bedform fields. The origin of the defects is thought to be linked to packets of hairpin vortex structures. These coherent turbulent structures give rise to entrainment corridors on the mobile bed, forming grain lineations that interact with the low-speed streaks generating an agglomeration of grains. Once a critical height of grains is reached, flow separation occurs over the new structure. Sediment will be eroded close from the reattachment point and deposited downstream creating a new defect. This new defect will thus induce formation of another defect and the process will continue, propagating downstream while the accumulations of grains quickly evolve into small bedforms.

Instantaneous Initiation

In general, the defect propagation theory plays a bigger role at low sediment transport rates since for high rates defects maybe washed away and bedforms generally initiated across the entire bed spontaneously. Venditti et al. (2005) report that instantaneous initiation begins with the formation of a cross-hatch pattern, which leads to chevron-shaped forms that migrate independently of the pattern structure. This chevron-like structure reorganizes to form the future crest lines of the bedforms. Venditti et al. (2006), based on the earlier model by Liu (1957), proposed that instantaneous initiation is a manifestation of an interfacial hydrodynamic instability of Kelvin-Helmholtz type between a highly active pseudofluid sediment layer and the fluid above it. In addition, Venditti et al. (2005)[2] imply that there is no linkage between the instantaneous initiation and coherent turbulent flow structures, since spatially- and temporally-random events should lock in place to generate the cross-hatch pattern. Moreover, there is no clear explanation of the effect of turbulence in the formation of bedforms since bedforms may also occur under laminar flows. It is important to note, that laminar-generated bedform studies used the temporally-averaged flow conditions to determine the degree of turbulence, indicating Reynolds number in the laminar regime. However, instantaneous process, such as burst and sweeps, which are infrequent at low Reynolds number but still present, can be the driving mechanisms to generate the bedforms. The generation of bedforms in laminar flows is still a topic of debate within the scientific community, since if true, it suggests that there should be other processes for defect development other than the one suggested by Best (1992).[1] This alternative model for bedform development at low sediment transport rates should explain the generation of defects and bedforms for cases where the flow is not turbulent.

Bedform Phase Diagrams

Phase or stability diagrams are defined as graphs that show the regimes of existence of one or more stable bed states. The stability of the bed can be defined when the bedform is in equilibrium and does not change in time for the same flow condition. This invariance over time must not be confused with a static morphology or frozen equilibrium; on the contrary, the bed moves and adjusts in a dynamic equilibrium with the flow and sediment transport for that particular condition. These phase diagrams are used for two main purposes: i) for prediction of bed states in a known flow and sediment transport condition, and, ii) as a tool for the reconstruction of paleoenvironments from a known bed state or sedimentary structure. Despite the great utility of such diagrams, they are very difficult to construct, making them either incomplete or very hard to interpret. This complexity lies in the number of variables needed to quantify the system.

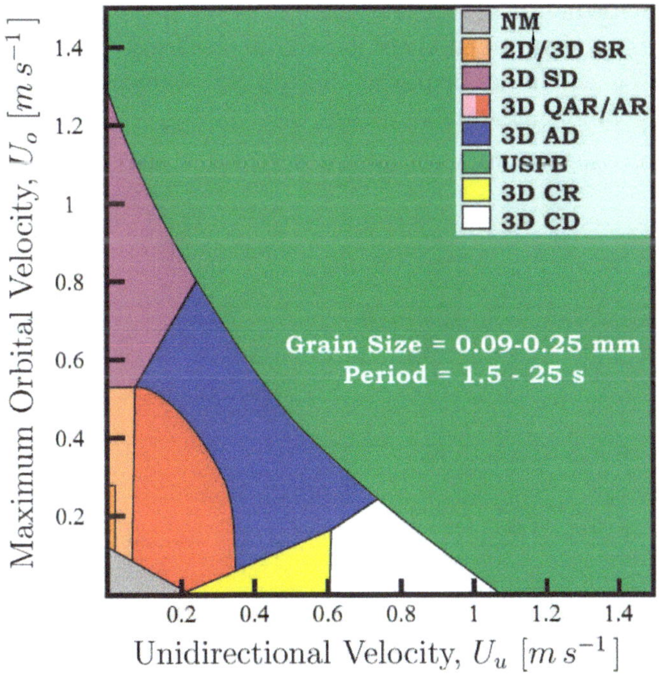

Dimensional phase diagram for combined flows. Relationships of combined-flow bed-phases stability fields in a plot of Oscillatory vs Unidirectional velocity.

Bedforms vs. Flow

Typical unidirectional bedforms represent a specific flow velocity, assuming typical sediments (sands and silts) and water depths, and a chart such as below can be used for interpreting depositional environments, with **increasing** water velocity going down the chart.

Flow Regime	**Bedform**	**Preservation Potential**	**Identification Tips**
Lower			
	Lower plane bed	High	Flat laminae, almost lack of current
	Ripple marks	High	Small, cm-scale undulations
	Sand waves	Medium to low	Rare, longer wavelength than ripples
	Dunes/Megaripples	High	Large, meter-scale ripples
Upper			
	Upper plane bed	High	Flat laminae, +/- aligned grains (parting lineations)
	Antidunes	Low	Water in phase with bedform, low angle, subtle laminae
	Pool and chute	Very low	Mostly erosional features

This chart is for general use, because changes in grain size and flow depth can change the bedform present and skip bedforms in certain scenarios. Bidirectional environments (e.g. tidal flats) produce similar bedforms, but the reworking the sediments and opposite directions of flow complicates the structures.

This bed form sequence can also be illustrated diagrammatically:

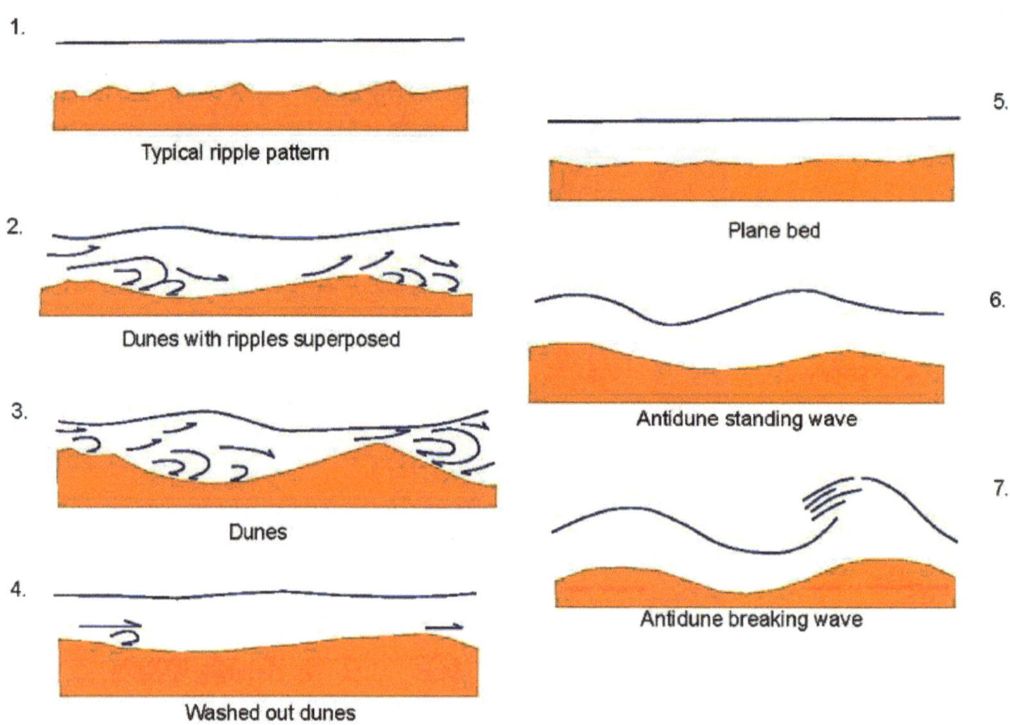

Bedforms formed in sand in channels under unidirectional flow. Numbers correspond broadly to increasing flow regime, i.e., increasing water flow velocity. Blue arrows show schematically flow lines in the water above the bed. Flow is always from left to right.

Types of Bedforms

Lower Plane Bed

"Lower plane bed" refers to the flat configuration the bed of a river that is produced in via low rates of sediment transport.

Upper Plane Bed

Parting lineation, from lower left to upper right; Kayenta Formation, Canyonlands National Park.

"Upper plane bed" features are flat and characterized by a unidirectional flow with high rates of sediment transport as both bed load and suspended load. Upper plane bed conditions can produce parting current lineations, which are typically subtle streaks on the bed surface due to the high energy flow.

Modern asymmetric ripples developed in sand on the floor of the Hunter River, New South Wales, Australia. Flow direction is from right to left.

Sinuous-crested dunes exposed at low tide in the Cornwallis River near Wolfville, Nova Scotia.

Ancient channel deposit in the Stellarton Formation (Pennsylvanian), Coalburn Pit, near Thorburn, Nova Scotia.

Sediment motion can create self-organized structures such as ripples, dunes, antidunes on the river or stream bed. These bedforms are often preserved in sedimentary rocks and can be used to estimate the direction and magnitude of the flow that deposited the sediment.

Surface Runoff

Overland flow can erode soil particles and transport them downslope. The erosion associated with overland flow may occur through different methods depending on meteorological and flow conditions.

- If the initial impact of rain droplets dislodges soil, the phenomenon is called rainsplash erosion.
- If overland flow is directly responsible for sediment entrainment but does not form gullies, it is called "sheet erosion".
- If the flow and the substrate permit channelization, gullies may form; this is termed "gully erosion".

Key Fluvial Depositional Environments

The major fluvial (river and stream) environments for deposition of sediments include:

- Deltas (arguably an intermediate environment between fluvial and marine)
- Point bars
- Alluvial fans
- Braided rivers
- Oxbow lakes
- Levees
- Waterfalls

River Delta

A river delta is a landform that forms from deposition of sediment carried by a river as the flow leaves its mouth and enters slower-moving or standing water. This occurs where a river enters an ocean, sea, estuary, lake, reservoir, or (more rarely) another river that cannot transport away the supplied sediment.

Despite a popular legend, this use of the word *delta* was not coined by Herodotus.

Sacramento (California) Delta at flood stage, early 2009.

Formation

River deltas form when a river carrying sediment reaches either (1) a body of water, such as a lake, ocean, or reservoir, (2) another river that cannot remove the sediment quickly enough to stop delta formation, or (3) an inland region where the water spreads out and deposits sediments. The tidal currents also cannot be too strong, as sediment would wash out into the water body faster than the river deposits it. Of course, the river must carry enough sediment to layer into deltas over time. The river's velocity decreases rapidly, causing it to deposit the majority, if not all, of its load. This alluvium builds up to form the river delta. When the flow enters the standing water, it is no longer confined to its channel and expands in width. This flow expansion results in a decrease in the flow velocity, which diminishes the ability of the flow to transport sediment. As a result, sediment drops out of the flow and deposits. Over time, this single channel builds a deltaic lobe (such as the bird's-foot of the Mississippi or Ural river deltas), pushing its mouth into the standing water. As the deltaic lobe advances, the gradient of the river channel becomes lower because the river channel is longer but has the same change in elevation.

As the slope of the river channel decreases, it becomes unstable for two reasons. First, gravity makes the water flow in the most direct course down slope. If the river breaches its natural levees (i.e., during a flood), it spills out onto a new course with a shorter route to the ocean, thereby obtaining a more stable steeper slope. Second, as its slope gets lower, the amount of shear stress on the bed decreases, which results in deposition of sediment within the channel and a rise in the channel bed relative to the floodplain. This makes it easier for the river to breach its levees and cut a new channel that enters the body of standing water at a steeper slope. Often when the channel does this, some of its flow remains in the abandoned channel. When these channel-switching events occur, a mature delta develops a distributary network.

Another way these distributary networks form is from deposition of mouth bars (mid-channel sand and/or gravel bars at the mouth of a river). When this mid-channel bar is deposited at the mouth of a river, the flow is routed around it. This results in additional deposition on the upstream end of the mouth-bar, which splits the river into two distributary channels. A good example of the result of this process is the Wax Lake Delta.

In both of these cases, depositional processes force redistribution of deposition from areas of high deposition to areas of low deposition. This results in the smoothing of the planform (or map-view) shape of the delta as the channels move across its surface and deposit sediment. Because the sediment is laid down in this fashion, the shape of these deltas approximates a fan. The more often the flow changes course, the shape develops as closer to an ideal fan, because more rapid changes in channel position results in more uniform deposition of sediment on the delta front. The Mississippi and Ural River deltas, with their bird's-feet, are examples of rivers that do not avulse often enough to form a symmetrical fan shape. Alluvial fan deltas, as seen by their name, avulse frequently and more closely approximate an ideal fan shape.

Types of Deltas

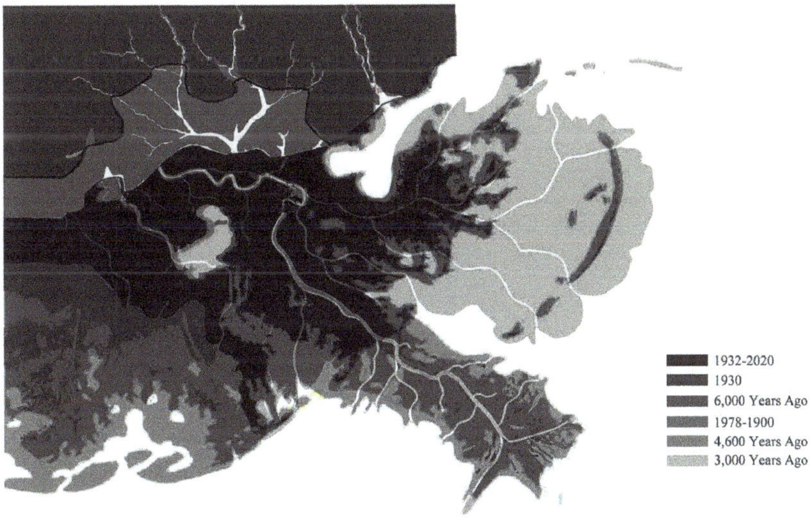

Lower Mississippi River land loss over time.

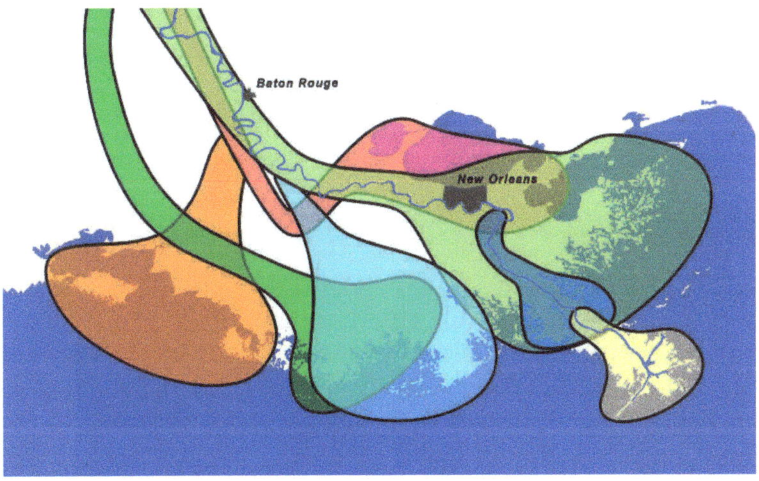

Delta lobe switching in the Mississippi Delta, 4600 yrs BP, 3500 yrs BP, 2800 yrs BP, 1000 yrs BP, 300 yrs BP, 500 yrs BP, current.

Deltas are typically classified according to the main control on deposition, which is usually either a river, waves, or tides. These controls have a large effect on the shape of the resulting delta.

Wave-Dominated Deltas

In wave dominated deltas, wave-driven sediment transport controls the shape of the delta, and much of the sediment emanating from the river mouth is deflected along the coast line.

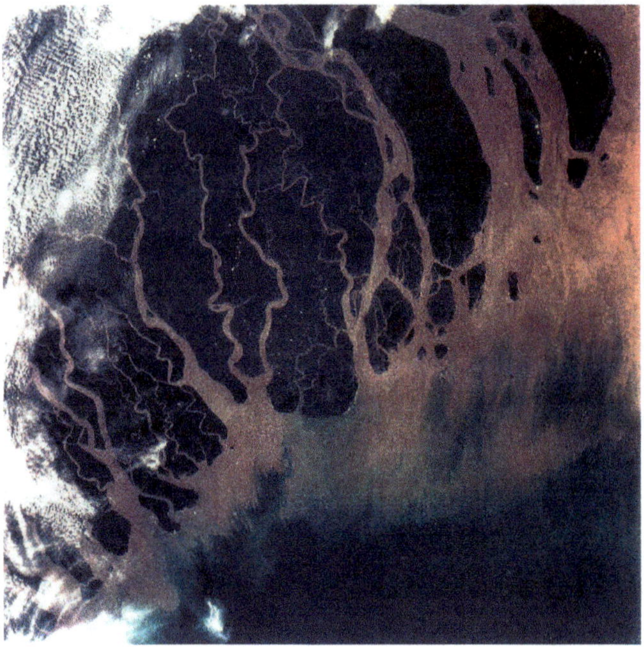

The Ganges Delta in India and Bangladesh is the largest delta in the world and it is also one of the most fertile regions in the world.

Tide-Dominated Deltas

Erosion is also an important control in tide dominated deltas, such as the Ganges Delta, which may be mainly submarine, with prominent sand bars and ridges. This tends to produce a "dendritic" structure. Tidal deltas behave differently from river- and wave-dominated deltas, which tend to have a few main distributaries. Once a wave- or river-distributary silts up, it is abandoned, and a new channel forms elsewhere. In a tidal delta, new distributaries are formed during times when there's a lot of water around – such as floods or storm surges. These distributaries slowly silt up at a pretty constant rate until they fizzle out.

Gilbert Deltas

A Gilbert delta (named after Grove Karl Gilbert) is a specific type of delta formed from coarse sediments, as opposed to gently-sloping muddy deltas such as that of the Mississippi. For example, a mountain river depositing sediment into a freshwater lake would form this kind of delta. While some authors describe both lacustrine and marine locations of Gilbert deltas, others note that their formation is more characteristic of the freshwater lakes, where it is easier for the river water to mix with the lakewater faster (as opposed to the case of a river falling into the sea or a salt lake, where less dense fresh water brought by the river stays on top longer).

G.K. Gilbert himself first described this type of delta on Lake Bonneville in 1885. Elsewhere, similar structures occur, for example, at the mouths of several creeks that flow into Okanagan Lake in British Columbia and forming prominent peninsulas at Naramata (49°35′30″N 119°35′30″W), Summerland (49°34′23″N 119°37′45″W), or Peachland (49°47′00″N 119°42′45″W)

Estuaries

Other rivers, particularly those on coasts with significant tidal range, do not form a delta but enter into the sea in the form of an estuary. Notable examples include the Saint Lawrence River and the Tagus estuary.

Inland Deltas

Okavango Delta.

In rare cases the river delta is located inside a large valley and is called an inverted river delta. Sometimes a river divides into multiple branches in an inland area, only to rejoin and continue to the sea. Such an area is called an *inland delta*, and often occurs on former lake beds. The Inner Niger Delta and Peace–Athabasca Delta are notable examples. The Amazon has also an inland delta before the island of Marajó.

In some cases, a river flowing into a flat arid area splits into channels that evaporate as it progresses into the desert. Okavango Delta in Botswana is one well-known example.

Sedimentary Structure

The formation of a delta is complicated, multiple, and cross-cutting over time, but in a simple delta three main types of bedding may be distinguished: the bottomset beds, foreset/frontset beds, and topset beds. This three part structure may be seen in small scale by crossbedding.

- The bottomset beds are created from the lightest suspended particles that settle farthest away from the active delta front, as the river flow diminishes into the standing body of water and loses energy. This suspended load is deposited by sediment gravity flow, creating a turbidite. These beds are laid down in horizontal layers and consist of the finest grain sizes.
- The foreset beds in turn are deposited in inclined layers over the bottomset beds as the active lobe advances. Foreset beds form the greater part of the bulk of a delta, (and also occur on the lee side of sand dunes). The sediment particles within foreset beds consist of larger and more variable sizes, and constitute the bed load that the river moves downstream by rolling and bouncing along the channel bottom. When the bed load reaches the edge of the delta front, it rolls over the edge, and is deposited in steeply dipping layers over the top of the existing bottomset beds. Under water, the slope of the outermost edge of the delta is created at the angle of repose of these sediments. As the foresets accumulate and advance, subaqueous landslides occur and readjust overall slope stability. The foreset slope, thus created and maintained, extends the delta lobe outward. In cross section, foresets typically lie in angled, parallel bands, and indicate stages and seasonal variations during the creation of the delta.
- The topset beds of an advancing delta are deposited in turn over the previously laid foresets, truncating or covering them. Topsets are nearly horizontal layers of smaller-sized sediment deposited on the top of the delta and form an extension of the landward alluvial plain. As the river channels meander laterally across the top of the delta, the river is lengthened and its gradient is reduced, causing the suspended load to settle out in nearly horizontal beds over the delta's top. Topset beds are subdivided into two regions: the upper delta plain and the lower delta plain. The upper delta plain is unaffected by the tide, while the boundary with the lower delta plain is defined by the upper limit of tidal influence.

Examples of Deltas

The Ganges/Brahmaputra combination delta which spans most of Bangladesh and empties into the Bay of Bengal, is the world's largest delta. Other rivers with notable deltas include the Nile River, the Fly River, the Godavari, the Krishna, the Kaveri, the Niger River, the Tigris-Euphrates, the Rhine, the Po, the Rhône, the Danube, the Ebro, the Volga, the Lena, the Indus, the Ayeyarwady (Irrawaddy), the Mekong, the Huanghe, the Yangtze, the Sacramento–San Joaquin, the Mississippi, the Orinoco, and the Paraná.

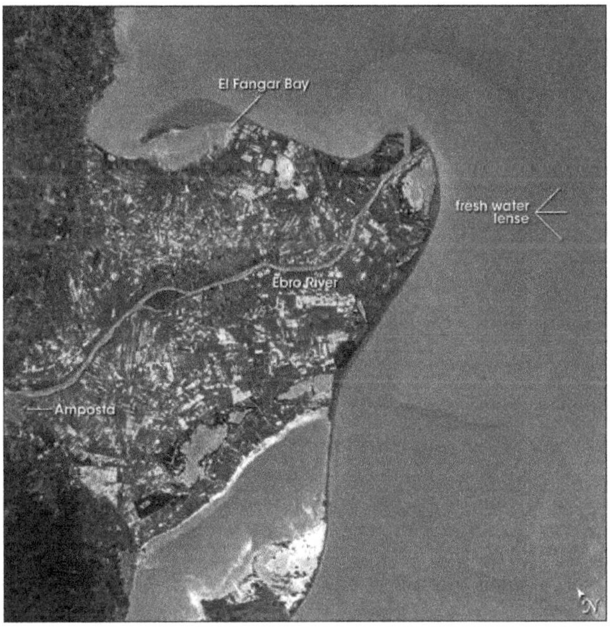

The Ebre River delta at the Mediterranean Sea.

The delta of the St. Clair River between the Canadian province of Ontario and the U.S. state of Michigan is the largest delta emptying into a body of fresh water.

Ecological Threats to Deltas

Human activities, including of water and the creation of dams for hydroelectric power or to create reservoirs can radically alter delta ecosystems. Dams block sedimentation, which can cause the delta to erode away. The use of water upstream can greatly increase salinity levels as less fresh water flows to meet the salty ocean water. While nearly all deltas have been impacted to some degree by humans, the Nile Delta and Colorado River Delta are some of the most extreme examples of the ecological devastation caused to deltas by damming and diversion of water.

Deltas in the Economy

Ancient deltas are a benefit to the economy due to their well sorted sand and gravel. Sand and gravel is often quarried from these old deltas and used in concrete for highways, buildings, sidewalks, and even landscaping. More than 1 billion tons of sand and gravel are produced in the United States alone. Not all sand and gravel quarries are former deltas, but for ones that are, a lot of the sorting is already done by the power of water.

As lowlands often adjacent to urban areas, deltas often comprise extensive industrial and commercial areas as well as agricultural land. These uses are often in conflict. The Fraser Delta in British Columbia, Canada, includes the Vancouver Airport and the Roberts Bank Superport and the Annacis Island industrial zone, and a mix of commercial, residential and agricultural land. Space is so limited in the Lower Mainland region, and in British Columbia in general, which is very mountainous, that the Agricultural Land Reserve was created to preserve agricultural land for food production.

Deltas on Mars

Researchers have found a number of examples of deltas that formed in Martian lakes. Finding deltas is a major sign that Mars once had large amounts of water. Deltas have been found over a wide geographical range. Below are pictures of a few.

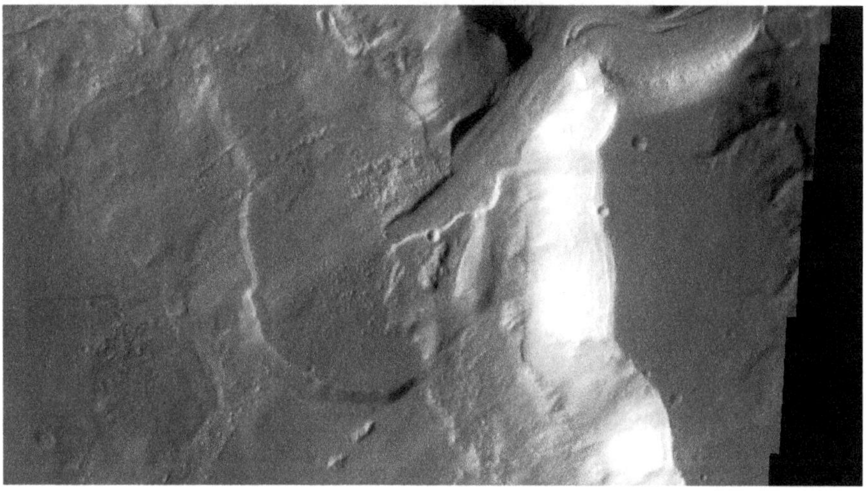

Delta in Ismenius Lacus quadrangle, as seen by THEMIS.

Delta in Lunae Palus quadrangle, as seen by THEMIS.

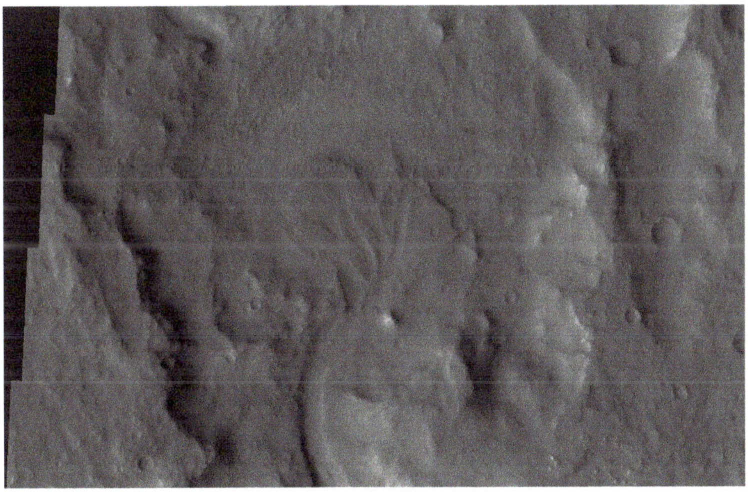

Delta in Margaritifer Sinus quadrangle as seen by THEMIS.

Probable delta in Eberswalde crater, as seen by Mars Global Surveyor. Image in Margaritifer Sinus quadrangle.

Point Bar

A point bar is a depositional feature made of alluvium that accumulates on the inside bend of streams and rivers below the slip-off slope. Point bars are found in abundance in mature or meandering streams. They are crescent-shaped and located on the inside of a stream bend, being very similar to, though often smaller than, towheads, or river islands.

Point bars are composed of sediment that is well sorted and typically reflects the overall capacity of the stream. They also have a very gentle *slope* and an elevation very close to water level. Since they are low-lying, they are often overtaken by floods and can accumulate driftwood and other debris during times of high water levels. Due to their near flat topography and the fact that the water speed is slow in the shallows of the point bar they are popular rest stops for boaters and rafters. However, camping on a point bar can be dangerous as a flash flood that raises the stream level by as little as a few inches (centimeters) can overwhelm a campsite in moments.

A point bar is an area of deposition whereas a cut bank is an area of erosion.

Point bars are formed as the secondary flow of the stream sweeps and rolls sand, gravel and small stones laterally across the floor of the stream and up the shallow sloping floor of the point bar.

Cut bank erosion and point bar deposition as seen on the Powder River in Montana.

Formation

Any fluid, including water in a stream, can only flow around a bend in vortex flow. In vortex flow the speed of the fluid is fastest where the radius is smallest, and slowest where the radius is greatest. (Tropical cyclones, tornadoes, and the spinning motion of water as it escapes down a drain are all visible examples of vortex flow.) In the case of water flowing around a bend in a stream the secondary flow in the boundary layer along the floor of the stream does not flow parallel to the banks of the stream but flows partly across the floor of the stream toward the inside of the stream (where the radius of curvature is smallest). This movement of the boundary layer is capable of sweeping and rolling loose particles including sand, gravel, small stones and other submerged objects along the floor of the stream toward the point bar.

Point bar at a river meander: the Cirque de la Madeleine in the Gorges de l'Ardèche, France.

This can be demonstrated at home. Partly fill a circular bowl or cup with water and sprinkle a little sand, rice or sugar into the water. Set the water in circular motion with a hand or spoon. The secondary flow will quickly sweep the solid particles into a neat pile in the center of the bowl or cup. The primary flow (the vortex) might be expected to sweep the solid particles to the perimeter of the bowl or cup, but instead the secondary flow along the floor of the bowl or cup sweeps the particles toward the center.

Where a stream is following a straight course the slower boundary layer along the floor of the stream is also following the same straight course. It sweeps and rolls sand, gravel and polished stones downstream, along the floor of the stream. However, as the stream enters a bend and vortex flow commences as the primary flow, a secondary flow also commences and flows partly across the floor of the stream toward the convex bank (the bank with the smaller radius). Sand, gravel and polished stones that have travelled with the stream for a great distance where the stream was following a straight course may finally come to rest in the point bar of the first stream bend.

Due to the circular path of a stream around a bend the surface of the water is slightly higher near the concave bank (the bank with the larger radius) than near the convex bank. This slight slope on the water surface of the stream causes a slightly greater water pressure on the floor of the stream near the concave bank than near the convex bank.

This pressure gradient drives the slower boundary layer across the floor of the stream toward the convex bank. The pressure gradient is capable of driving the boundary layer up the shallow sloping floor of the point bar, causing sand, gravel and polished stones to be swept and rolled up-hill.

The concave bank is often a cut bank and an area of erosion. The eroded material is swept and rolled across the floor of the stream by the secondary flow and may be deposited on the point bar only a small distance downstream from its original location in the concave bank.

The point bar typically has a gently sloping floor with shallow water. The shallow water is mostly the accumulated boundary layer and does not have a fast speed. However, in the deepest parts of the stream where the stream is flowing freely, vortex flow prevails and the stream is flowing fastest where the radius of the bend is smallest, and slowest where the radius is greatest. The shallows around the point bar can become treacherous when the stream is rising. As the water depth increases over the shallows of the point bar, the vortex flow can extend closer toward the convex bank and the water speed at any point can increase dramatically in response to only a small increase in water depth.

Fallacy Regarding Formation of Point Bars

An enduring fallacy exists regarding the formation of point bars and oxbow lakes. The fallacy suggests they are formed by the deposition of a stream's suspended load as the velocity and energy of the stream *decreases* in the bend. This fallacy relies on the erroneous notion that the speed of the water is slowest on the inside of the bend (where the radius is smallest) and fastest on the outside of the bend (where the radius is greatest).

If it were true that, around a bend in a stream, the difference between the speed of the water near one bank and the other was sufficient to cause deposition of suspended solids on one bank but not the other, the deposition would occur near the concave bank rather than the convex bank because vortex flow is slowest where the radius of curvature is greatest. Any fluid, including water in a stream, can only flow around a bend in vortex flow.

All point bars typically have a gently sloping floor with shallow water. The shallow depth of the water, and the fact that it is an accumulation of the boundary layer, prevent the water from reaching fast speed over the point bar. It is probably this observation which led early geographers to believe the slowest part of the stream is where the radius is smallest.

In a slow-flowing stream or river, the difference in speed between one bank and the other is not sufficient to provide a credible explanation as to why all the deposition occurs on one bank, and none on the other. Similarly, the fallacy has no explanation as to why all the deposition occurs at a stream bend, and little or none occurs where the stream is following a straight course. (The speed of water in a stream does not slow just because the stream enters a bend.)

In a mature, meandering stream or river the water speed is slow, turbulence is low, and the water is not capable of holding coarse sand and gravel in suspension. In contrast, point bars comprise coarse sand, gravel, polished stones and other submerged objects.

These materials have not been carried in suspension and then dropped on the point bar – they have been swept and rolled into place by the secondary flow that exists across the floor of every stream in the vicinity of a stream bend.

Braided River

A braided river (or braided channel) is one of a number of channel types and has a channel that consists of a network of small channels separated by small and often temporary islands called braid bars or, in British usage, *aits* or *eyots*. Braided streams occur in rivers with high slope and/or large sediment load. Braided channels are also typical of environments that dramatically decrease channel depth, and consequently channel velocity, such as river deltas, alluvial fans and peneplains.

The Waimakariri River is braided over most of its course.

Formation

Braided rivers, as distinct from meandering rivers, occur when a threshold level of sediment load or slope is reached whilst a steep gradient is also maintained. Geologically speaking, an increase in sediment load will over time increase in the slope of the river, so these two conditions can be considered synonymous; and, consequently, a variation of slope can model a variation in sediment load. A threshold slope was experimentally determined to be 0.016 (ft/ft) for a 0.15 cu ft/s (0.0042 m^3/s) stream with poorly sorted coarse sand. Any slope over this threshold created a braided stream, while any slope under the threshold created a meandering stream or— for very low slopes—a straight channel. So the main controlling factor on river development is the amount of sediment that the river carries; once a given system crosses a threshold value for sediment load, it will convert from a meandering system to a braided system. Also important to channel development is the proportion of suspended load sediment to bed load. An increase in suspended sediment allowed for the deposition of fine erosion-resistant material on the inside of a curve, which accentuated the curve and in some instances caused a river to shift from a braided to a meandering profile. The channels and braid bars are usually highly mobile, with the river layout often changing significantly during flood events. Channels move sideways via differential velocity: On the outside of a curve, deeper, swift water picks up sediment (usually gravel or larger stones), which is re-deposited in slow-moving water on the inside of a bend.

The White River in the U.S. state of Washington transports a large sediment load from the Emmons Glacier of Mount Rainier, a young, rapidly eroding volcano.

The braided channels may flow within an area defined by relatively stable banks or may occupy an entire valley floor. The Rakaia River in Canterbury, New Zealand has cut a channel 100 meters wide into the surrounding plains; this river transports sediment to a lagoon located on the river-coast interface.

Conditions associated with braided channel formation include:

- an abundant supply of sediment
- high stream gradient
- rapid and frequent variations in water discharge
- erodible banks
- a steep channel gradient

However, the critical factor that determines whether a stream will meander or braid is bank erodibility. A stream with cohesive banks that are resistant to erosion will form narrow, deep, meandering channels, whereas a stream with highly erodible banks will form wide, shallow channels, sustaining helical flow and resulting in the formation of braided channels.

Examples

Extensive braided river systems are found in Alaska, Canada, New Zealand's South Island, and the Himalayas, which all contain young, rapidly eroding mountains.

- The enormous Brahmaputra-Jamuna River in Asia is a classic example of a braided river.
- Braided river systems are present in Africa, for example in the Touat Valley.
- A notable example of a large braided stream in the contiguous United States is the Platte River in central and western Nebraska. The sediment of the arid Great Plains is augmented by the presence of the nearby Sandhills region north of the river.
- A portion of the lower Yellow River takes a braided form.
- The Sewanee Conglomerate, a Pennsylvanian coarse sandstone and conglomerate unit present on the Cumberland Plateau near the University of the South, may have been deposited by an ancient braided and meandering river that once existed in the eastern United States. Others have interpreted the depositional environment for this unit as a tidal delta.

Notable braided rivers in Europe:

- Italy
    - Tagliamento (Northeastern Italy)
    - Piave (river)
    - Brenta (river)
    - Cellina
    - Meduna
    - Fella
    - Magra
- Narew (Poland and Belarus)

Distinction Between Braided Rivers and Anastomosing Rivers

Anastomosing rivers or streams are similar to braided rivers in that they consist of multiple interweaving channels. However, anastomosing rivers typically consist of a network of low-gradient, narrow, deep channels with stable banks, in contrast to braided rivers, which form on steeper gradients and display less bank stability.

Oxbow Lake

An oxbow lake is a U-shaped body of water that forms when a wide meander from the main stem of a river is cut off, creating a free-standing body of water. This landform is so named for its distinctive curved shape, resembling the bow pin of an oxbow. In Australia, an oxbow lake is known as a *billabong,* from the indigenous language Wiradjuri. In south Texas, oxbows left by the Rio Grande are called *resacas.*

The word "oxbow" can also refer to a U-shaped bend in a river or stream, whether or not it is cut off from the main stream.

This picture of the Nowitna River, Alaska shows two oxbow lakes - a short one at the bottom of the picture and a longer, more curved one at the middle-right. Furthermore, it can be clearly seen how a third oxbow lake is in the making. The isthmus or bank in the center of the most prominent meander is very narrow - much narrower than the width of the river. Eventually the two sections of river on either side of the isthmus will break through and create a new, straighter course. A new river bank then starts to accumulate, sealing off the meander and leaving another oxbow lake.

Geology

An oxbow lake forms when a river creates a meander, due to the river's eroding the bank through hydraulic action, abrasion and erosion. After a long period of time, the meander becomes very curved, and eventually the neck of the meander becomes narrower and the river cuts through the neck during a flood, cutting off the meander and forming an oxbow lake.

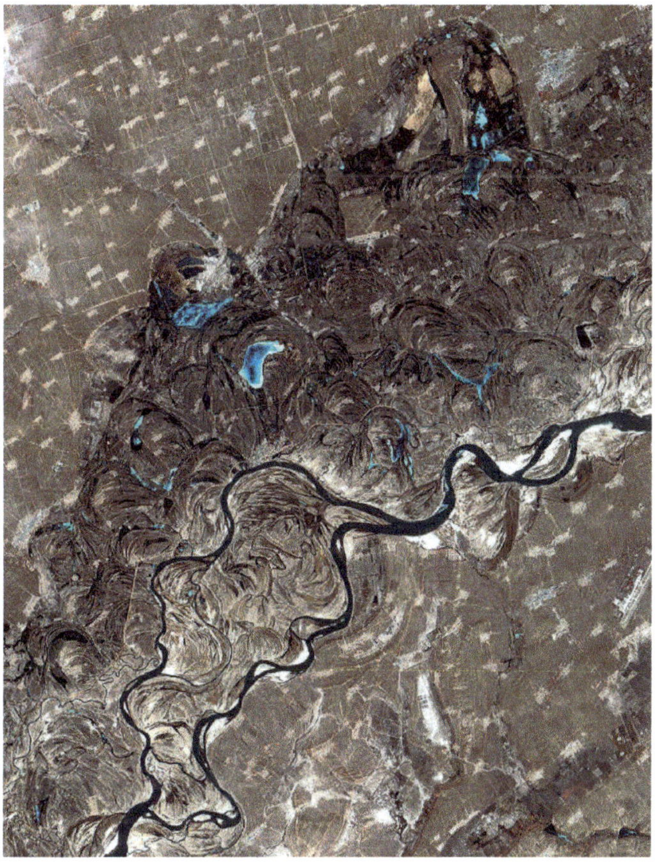

The swirls and curves of the previous courses of the Songhua River in northeast China can easily be seen in this satellite photo

When a river reaches a low-lying plain, often in its final course to the sea or a lake, it meanders widely. In the vicinity of a river bend, deposition occurs on the convex bank (the bank with the smaller radius). In contrast, both lateral erosion and undercutting occur on the cut bank or concave bank (the bank with the greater radius.) Continuous deposition on the convex bank and erosion of the concave bank of a meandering river cause the formation of a very pronounced meander with two concave banks getting closer. The narrow neck of land between the two neighboring concave banks is finally cut through, either by lateral erosion of the two concave banks or by the strong currents of a flood. When this happens, a new straighter river channel is created and an abandoned meander loop, called a cutoff, is formed. When deposition finally seals off the cutoff from the river channel, an oxbow lake is formed. This process can occur over a time scale from a few years to several decades and may sometimes become essentially static.

Gathering of erosion products near the concave bank and transporting them to the convex bank is the work of the secondary flow across the floor of the river in the vicinity of a river bend. The process of deposition of silt, sand and gravel on the convex bank is clearly illustrated in point bars.

River flood plains that contain rivers with a highly sinuous platform are populated by longer oxbow lakes than those with low sinuosity. This is because rivers with high sinuosity have larger meanders, and greater opportunity for longer lakes to form. Rivers with lower sinuosity are characterized by fewer cutoffs and shorter oxbow lakes due to the shorter distance of their meanders.

The effect of the secondary flow can be demonstrated using a circular bowl. Partly fill the bowl with water and sprinkle dense particles such as sand or rice into the bowl. Set the water into circular motion with one hand or a spoon. The dense particles quickly sweep into a neat pile in the center of the bowl. This is the mechanism that leads to the formation of point bars and contributes to the formation of oxbow lakes. The primary flow of water in the bowl is circular and the streamlines are concentric with the side of the bowl. However, the secondary flow of the boundary layer across the floor of the bowl is inward toward the center. The primary flow might be expected to fling the dense particles to the perimeter of the bowl, but instead the secondary flow sweeps the particles toward the center.

The curved path of a river around a bend makes the water's surface slightly higher on the outside of the bend than on the inside. As a result, at any elevation within the river, water pressure is slightly greater near the outside of the bend than on the inside. A pressure gradient toward the convex bank provides the centripetal force necessary for each parcel of water to follow its curved path.

The boundary layer that flows along the river floor does not move fast enough to balance the pressure gradient laterally across the river. It responds to this pressure gradient, and its velocity is partly downstream and partly across the river toward the convex bank. As it flows along the floor of the river, it sweeps loose material toward the convex bank. This flow of the boundary layer is significantly different from the speed and direction of the primary flow of the river, and is part of the river's secondary flow.

Early stages of formation of coastal plain oxbow lake in the Gower Peninsula of southwest Wales.

A Horseshoe or oxbow lake near Hughes, Arkansas. The bulges in the border reflect changes in the course of the river; when the river shifted its course and cut off the former channel, the border remained unchanged.

When a fluid follows a curved path, such as around a circular bowl, around a bend in a river or in a tropical cyclone, the flow is described as vortex flow: the fastest speed occurs where the radius is smallest, and the slowest speed occurs where the radius is greatest. The higher fluid pressure and slower speed where the radius is greater, and the lower pressure and faster speed where the radius is smaller, are all consistent with Bernoulli's principle.

Notable examples

- Bole and Burton Round in West Burton, Nottinghamshire, England are a good example of previous lakes in a close proximity to one another.
- Carter Lake, Iowa was created after severe flooding in 1877 led to the river shifting approximately 1.25 mi to the southeast.
- Cuckmere Haven in Sussex, England contains a widely meandering river with many oxbow lakes, often referred to in physical geography textbooks.
- Half Moon Lake in downtown Eau Claire, Wisconsin was formed due to a shift in the course of the Chippewa River, which now flows immediately to the south.
- The towns of Horseshoe Lake, Arkansas and Lake Village, Arkansas are named after horseshoe-shaped oxbow lakes on which each city is located.
- Kanwar Lake Bird Sanctuary, India contains rare and endangered migratory birds and is one of Asia's largest oxbow lakes.
- The Oxbow, a 2.5-mile (4.0 km) bend in the Connecticut River, is disconnected at one end.

- The Reelfoot Lake in west Tennessee is an oxbow lake formed when the Mississippi River took a new channel following the New Madrid earthquake of 1811–1812. There are many oxbow lakes alongside the Mississippi River and its tributaries. The largest oxbow lake in North America, Lake Chicot (located near Lake Village, Arkansas), was originally part of the Mississippi River.

Artificial Oxbow Lakes

Oxbow lakes may be formed when a river channel is straightened artificially to improve navigation or for flood alleviation. This occurred notably on the upper Rhine in Germany in the nineteenth century.

An example of an entirely artificial waterway with oxbows is the Oxford Canal in England. When originally constructed, it had a very meandering course, following the contours of the land, but the northern part of the canal was straightened out between 1829 and 1834, reducing its length from 91 to 77 and a half miles (approximately) and creating a number of oxbow-shaped sections isolated from the new course.

REFERENCES

1. K.K.E. Neuendorf, J.P. Mehl, Jr., and J.A. Jackson, eds., 2005, Glossary of Geology. American Geological Institute, Alexandria, Virginia. 800 pp.
2. Wilson, W.E. & Moore, J.E. 2003. Glossary of Hydrology, American Geological Institute, Springer, 248pp.
3. Charlton, Ro (2008). Fundamentals of fluvial geomorphology. London: Rutledge. p. 234. ISBN 978-0-415-33454-9.
4. Wohl, Ellen (2014). Rivers in the Landscape: Science and Management. Wiley-Blackwell. p. 330. ISBN 1118414896.
5. Mathur, Anuradha; Dilip da Cunha (2001). Mississippi Floods: Designing a Shifting Landscape. New Haven, CT: Yale University Press. ISBN 0-300-08430-7
6. Dill, William A. (1990). Inland fisheries of Europe. Rome, Italy: UN Food and Agriculture Organization. ISBN 92-5-102999-7. http://www.fao.org/docrep/009/t0377e/t0377e00.htm
7. Prothero. sedimentary geology.
8. (PDF) https://scholarworks.iu.edu/dspace/bitstream/handle/2022/205/B04.pdf?sequence=1. Missing or empty |title= (help)
9. Prothero, Donald R., and Fred Schwab. Sedimentary Geology. Macmillan, 2004. APA
10. 1950. Web. <https://scholarworks.iu.edu/dspace/bitstream/handle/2022/205/B04.pdf?sequence=1>.
11. Krzic M., K. Watson, S. Grand, A. Bomke, S. Smith, S. Dyanatkar, C. Crowley, and G. Lascu. 2010. Soil Formation and Parent Material. The University of British Columbia, Vancouver, Thompson Rivers University, Kamloops, and

Agriculture and Agri-Food Canada, Summerland.

12. Erosion". Encyclopedia Britannica. 2015-12-03.
13. Cheraghi, M., S. Jomaa, G. C. Sander, and D. A. Barry (2016), Hysteretic sediment fluxes in rainfall-driven soil erosion: Particle size effects, Water Resour. Res., 52, doi:10.1002/ 2016WR019314 file [http://onlinelibrary.wiley.com/doi/10.1002/2016WR019314/full link]
14. Hallet, Bernard (1981). «Glacial Abrasion and Sliding: Their Dependence on the Debris Concentration In Basal Ice». Annals of Glaciology. 2 (1): 23–28. Bibcode:1981AnGla...2...23H. doi:10.3189/172756481794352487. ISSN 0260-3055.
15. Sklar, Leonard S.; Dietrich, William E. (2004). «A mechanistic model for river incision into bedrock by saltating bed load» (PDF). Water Resources Research. 40 (6): W06301. Bibcode:2004WRR....40.6301S. doi:10.1029/2003WR002496. ISSN 0043-1397.
16. Blanco-Canqui, Humberto; Rattan, Lal (2008). «Soil and water conservation». Principles of soil conservation and management. Dordrecht: Springer. pp. 1–20. ISBN 9781402087097.
17. Toy, Terrence J.; Foster, George R.; Renard, Kenneth G. (2002). Soil erosion : processes, prediction, measurement, and control. New York, NY: Wiley. ISBN 9780471383697.
18. Julien, Pierre Y. (2010). Erosion and Sedimentation. Cambridge University Press. p. 1. ISBN 978-0-521-53737-7.
19. Zachar, Dušan (1982). «Classification of soil erosion». Soil Erosion. Vol. 10. Elsevier. p. 48. ISBN 978-0-444-99725-8.
20. See Figure 1 in Obreschkow, D.; Dorsaz, N.; Kobel, P.; De Bosset, A.; Tinguely, M.; Field, J.; Farhat, M. (2011). «Confined Shocks inside Isolated Liquid Volumes - A New Path of Erosion?». Physics of Fluids. 23 (10): 101702. arXiv:1109.3175Freely accessible. Bibcode:2011PhFl...23j1702O. doi:10.1063/1.3647583.
21. Cheraghi, M., S. Jomaa, G. C. Sander, and D. A. Barry (2016), Hysteretic sediment fluxes in rainfall-driven soil erosion: Particle size effects, Water Resour. Res., 52, doi:10.1002/ 2016WR019314
22. Food and Agriculture Organization (1965). «Types of erosion damage». Soil Erosion by Water: Some Measures for Its Control on Cultivated Lands. United Nations. pp. 23–25. ISBN 978-92-5-100474-6.
23. Nearing, M.A.; Norton, L.D.; Bulgakov, D.A.; Larionov, G.A.; West, L.T.; Dontsova, K.M. (1997). «Hydraulics and erosion in eroding rills». Water Resources Research. 33 (4): 865–876. Bibcode:1997WRR....33..865N. doi:10.1029/97wr00013.
24. Boardman, John; Poesen, Jean, eds. (2007). Soil Erosion in Europe. Chichester: John Wiley & Sons. ISBN 9780470859117.
25. J. Poesen, L. Vandekerckhove, J. Nachtergaele, D. Oostwoud Wijdenes, G. Verstraeten, B. Can Wesemael (2002). «Gully erosion in dryland environments».

In Bull, Louise J. & Kirby, M.J. Dryland Rivers: Hydrology and Geomorphology of Semi-Arid Channels. John Wiley & Sons. pp. 229–262. ISBN 978-0-471-49123-1.

26. Borah, Deva K. et al. (2008). «Watershed sediment yield». In Garcia, Marcelo H. Sedimentation Engineering: Processes, Measurements, Modeling, and Practice. ASCE Publishing. p. 828. ISBN 978-0-7844-0814-8.
27. Ritter, Michael E. (2006) «Geologic Work of Streams» The Physical Environment: an Introduction to Physical Geography University of Wisconsin, OCLC 79006225
28. Nancy D. Gordon (2004-06-01). «Erosion and Scour». Stream hydrology: an introduction for ecologists. ISBN 978-0-470-84357-4.
29. «Thermal Erosion». NSIDC Glossary. National Snow and Ice Data Center. Archived from the original on 2010-11-18. Retrieved 21 December 2009. [permanent dead link]
30. Costard, F.; Dupeyrat, L.; Gautier, E.; Carey-Gailhardis, E. (2003). «Fluvial thermal erosion investigations along a rapidly eroding river bank: application to the Lena River (central Siberia)». Earth Surface Processes and Landforms. 28 (12): 1349–1359. Bibcode:2003ESPL...28.1349C. doi:10.1002/esp.592.
31. Jones, B.M.; Hinkel, K.M.; Arp, C.D.; Eisner, W.R. (2008). «Modern Erosion Rates and Loss of Coastal Features and Sites, Beaufort Sea Coastline, Alaska». Arctic. Arctic Institute of North America. 61 (4): 361–372. doi:10.14430/arctic44.
32. Glynn, Peter W. «Bioerosion and coral-reef growth: a dynamic balance.» Life and death of coral reefs (1997): 68-95.
33. Dixon, John C.; Thorn, Colin E. (2005). «Chemical weathering and landscape development in mid-latitude alpine environments». Geomorphology. 67 (1–2): 127–145. Bibcode:2005Geomo..67..127D. doi:10.1016/j.geomorph.2004.07.009. Retrieved 22 November 2015.
34. Thomson, Stuart N.; Brandon, Mark T.; Tomkin, Jonathan H.; Reiners, Peter W.; Vásquez, Cristián; Wilson, Nathaniel J. (2010). «Glaciation as a destructive and constructive control on mountain building». Nature. 467 (7313): 313–317. Bibcode:2010Natur.467..313T. doi:10.1038/nature09365. PMID 20844534.
35. Tomkin, J. H.; Roe, G. H. (2007). «Climate and tectonic controls on glaciated critical-taper orogens». Earth Planet. Sci. Lett. 262 (3–4): 385–397. Bibcode:2007E&PSL.262..385T. doi:10.1016/j.epsl.2007.07.040.
36. Mitchell, S. G. & Montgomery, D. R. Influence of a glacial buzzsaw on the height and morphology of the Cascade Range in central Washington State, USA. Quat. Res. 65, 96–107 (2006)
37. See, for example: Alt, David (2001). Glacial Lake Missoula & its Humongous Floods. Mountain Press. ISBN 978-0-87842-415-3.
38. Zheng, Xiaojing & Huang, Ning (2009). Mechanics of Wind-Blown Sand Movements. Springer. pp. 7–8. ISBN 978-3-540-88253-4.
39. Cornelis, Wim S. (2006). «Hydroclimatology of wind erosion in arid and semi-arid environments». In D›Odorico, Paolo & Porporato, Amilcare. Dryland

Ecohydrology. Springer. p. 141. ISBN 978-1-4020-4261-4.

40. Blanco-Canqui, Humberto; Rattan, Lal (2008). «Wind erosion». Principles of soil conservation and management. Dordrecht: Springer. pp. 54–80. ISBN 9781402087097.

41. Balba, A. Monem (1995). «Desertification: Wind erosion». Management of Problem Soils in Arid Ecosystems. CRC Press. p. 214. ISBN 978-0-87371-811-0.

42. Wiggs, Giles F.S. (2011). «Geomorphological hazards in drylands». In Thomas, David S.G. Arid Zone Geomorphology: Process, Form and Change in Drylands. John Wiley & Sons. p. 588. ISBN 978-0-470-71076-0.

43. Van Beek, Rens (2008). «Hillside processes: mass wasting, slope stability, and erosion». In Norris, Joanne E. et al. Slope Stability and Erosion Control: Ecotechnological Solutions. Springer. ISBN 978-1-4020-6675-7.

44. Gray, Donald H. & Sotir, Robbin B. (1996). «Surficial erosion and mass movement». Biotechnical and Soil Bioengineering Slope Stabilization: A Practical Guide for Erosion Control. John Wiley & Sons. p. 20. ISBN 978-0-471-04978-4.

45. Nichols, Gary (2009). Sedimentology and Stratigraphy. John Wiley & Sons. ISBN 978-1-4051-9379-5.

46. Blanco-Canqui, Humberto; Rattan, Lal (2008). «Water erosion». Principles of soil conservation and management. Dordrecht: Springer. pp. 21–53. ISBN 9781402087097.

47. Styczen, M.E. & Morgan, R.P.C. (1995). «Engineering properties of vegetation». In Morgan, R.P.C. & Rickson, R. Jane. Slope Stabilization and Erosion Control: A Bioengineering Approach. Taylor & Francis. ISBN 978-0-419-15630-7.

48. Whisenant, Steve G. (2008). «Terrestrial systems». In Perrow Michael R. & Davy, Anthony J. Handbook of Ecological Restoration: Principles of Restoration. Cambridge University Press. p. 89. ISBN 978-0-521-04983-2.

49. Wainwright, John & Brazier, Richard E. (2011). «Slope systems». In Thomas, David S.G. Arid Zone Geomorphology: Process, Form and Change in Drylands. John Wiley & Sons. ISBN 978-0-470-71076-0.

50. Burbank, Douglas W.; Anderson, Robert S. (2011). «Tectonic and surface uplift rates». Tectonic Geomorphology. John Wiley & Sons. pp. 270–271. ISBN 978-1-4443-4504-9.

51. Zeitler, P. K. et al. (2001), Erosion, Himalayan Geodynamics, and the Geomorphology of Metamorphism, GSA Today, 11, 4–9.

52. Pitman, W. C.; Golovchenko, X. (1991). «The effect of sea level changes on the morphology of mountain belts». Journal of Geophysical Research: Solid Earth. 96 (B4): 6879–6891. doi:10.1029/91JB00250. ISSN 0148-0227.

53. Beckinsale, Robert P.; Chorley, Richard J. (2003) [1991]. «Chapter Seven: American Polycyclic Geomorphology». The History of the Study of Landforms. Volume Three. Taylor & Francis e-Library. pp. 235–236.

54. Dewey, J.F.; Ryan, P.D.; Andersen, T.B. (1993). «Orogenic uplift and collapse, crustal thickness, fabrics and metamorphic phase changes: the role of eclogites».

Geological Society, London, Special Publications. 76 (1): 325–343.

55. Orlov, S.Yu.; Kuznetsov, N.B.; Miller, E.D.; Soboleva, A.A.; Udoratina, O.V. (2011). «Age Constraints for the Pre–Uralide–Timanide Orogenic Event Inferred from the Study of Detrital Zircons» (PDF). Doklady Earth Sciences. 440 (1): 1216–1221. Bibcode:2011DokES.440.1216O. doi:10.1134/s1028334x11090078. Retrieved 22 September 2015.
56. Lupia-Palmieri, Elvidio (2004). «Erosion». In Goudie, A.S. Encyclopedia of Geomorphology. p. 336.
57. Alexander, Earl B. (2014). Soils in natural landscapes. CRC Press. p. 108. ISBN 978-1-4665-9436-4.
58. Blanco, Humberto & Lal, Rattan (2010). «Soil and water conservation». Principles of Soil Conse

CHAPTER 6

Glacial Environments

Chapter Contents

INTRODUCTION

Glacial environments represent very harsh but beautiful environments and are located in the polar regions or in high mountainous regions. Glaciers are made of solid compacted snow that forms ice. There are several types of glacier, including cirques, valley glaciers, piedmonts and ice sheets. Glaciers change size throughout the year as well as over longer periods of time.

Glaciers can be viewed as a system with its balance between inputs, storages and outputs. Inputs include the accumulation of snow and debris. The glacier is also subject to heating processes and carries meltwater. The area of inputs is called the zone of accumulation. The main storage is ice but also debris that becomes locked inside the glacier and transported as moraine. There are different types of moraine, sub glacial moraine, englacial morain and medial moraine. The output is mainly meltwater and calfing sections of ice. This meltwater transports debris from within the glacier. This area of output is called the zone of ablation. The regime of a glacier refers to whether the glacier is in a stage of advance or in retreat. If accumulation exceeds ablation then the glacier will advance. If the ablation exceeds accumulation then the glacier will retreat.

TILL

Till or glacial till is unsorted glacial sediment.

Till is derived from the erosion and entrainment of material by the moving ice of a glacier. It is deposited some distance down-ice to form terminal, lateral, medial and ground moraines.

Till is classified into primary deposits, laid down directly by glaciers, and secondary deposits, reworked by fluvial transport and other processes.

Closeup of glacial till. Note that the larger grains (pebbles and gravel) in the till are completely surrounded by the matrix of finer material (silt and sand), and this characteristic, known as matrix support, is diagnostic of till.

Glacial till with tufts of grass.

Till after avalanche, Norway.

Processes

Glacial drift is the coarsely graded and extremely heterogeneous sediment of a glacier; till is the part of glacial drift deposited directly by the glacier. Its content may vary from clays to mixtures of clay, sand, gravel, and boulders. This material is mostly derived from the subglacial erosion and entrainment by the moving ice of the glaciers of previously available unconsolidated sediments. Bedrock can also be eroded through the action of glacial plucking and abrasion and the resulting clasts of various sizes will be incorporated to the glacier's bed.

Eventually, the sedimentary assemblage forming this bed will be abandoned some distance down-ice from its various sources. This is the process of glacial till deposition. When this deposition occurs at the base of the moving ice of a glacier, the sediment is called lodgement till. Rarely, eroded unconsolidated sediments can be preserved in the till along with their original sedimentary structures. More commonly, these sediments lose their original structure through the mixture processes associated with subglacial transport and they solely contribute to form the more or less uniform matrix of the till.

Till is deposited at the terminal moraine, along the lateral and medial moraines and in the ground moraine of a glacier. As a glacier melts, especially a continental glacier, large amounts of till are washed away and deposited as outwash in sandurs by the rivers flowing from the glacier, and as varves (annual layers) in any proglacial lakes which may form.

Till may contain detectable concentrations of gems or other valuable ore minerals picked up by the glacier during its advance, for example the diamonds found in the U.S. states of Wisconsin, Indiana, and in Canada. Prospectors use trace minerals in tills as clues to follow the glacier upstream to find kimberlite diamond deposits and other types of ore deposits.

Tillite

In cases where till has been indurated or lithified by subsequent burial into solid rock, it is known as the sedimentary rock *tillite*. Matching beds of ancient tillites on opposite sides of the south Atlantic Ocean provided early evidence for continental drift. The same tillites also provide some support to the Precambrian Snowball Earth glaciation event hypothesis.

Types of Till

There are various types of classifying tills:

- primary deposits – laid down directly by glacier action
- secondary deposits – reworked by fluvial transport, erosion, etc.

Traditionally (e.g. Dreimanis, 1988) a further set of divisions has been made to primary deposits, based upon the method of deposition.

- Lodgement tills – sediment which has been deposited by plastering of glacial debris from a sliding glacier bed.
- Deformation tills – Sediment which has been disaggregated and (usually) homogenized by shearing in the sub glacial deformed layer.
- Melt out tills – Released by melting of stagnant or slowly moving debris-rich glacier ice and deposited without subsequent transport or deformation. Split up into sub glacial melt out till (melting of debris rich ice at the bottom of the glacier) and supraglacial melt-out till (melting of ice on the glacier surface).
- Sublimation till – similar to melt out till, except the ice is lost through sublimation rather than melt. Often occurs only in extremely cold and arid conditions, mainly in Antarctica.

Van der Meer et al. 2003 have suggested that these till classifications are outdated and should instead be replaced with only one classification, that of deformation till. The reasons behind this are largely down to the difficulties in accurately classifying different tills, which are often based on inferences of the physical setting of the till rather than detailed analysis of the till fabric or particle size.

GLACIER

A glacier is a large body of ice and snow. It forms because the snow in an area does not all melt in summer. Each winter, more snow is added. The weight of all the snow creates pressure. This pressure turns the lower parts of the snow into ice. After this happens for many years, the glacier will start growing large. It becomes so heavy that gravity causes the ice to move. It flows downwards like water but very slowly. A glacier only moves about 50 meters (160 ft) per year. New snowfalls replace the parts that flow away.

Glaciers are the largest sources of fresh water on Earth. The largest bodies of salt water are the Oceans

The Baltoro Glacier in the Karakoram Mountains. 62 kilometers (39 mi) in length, one of the longest alpine-type glaciers.

Creation

Icebergs breaking off glaciers at Cape York, Greenland.

Glaciers will only form in places that are cold enough and get enough snow over time. This can take a long time. If often take tens or hundreds of years for a glacier to form. There are two kinds of glaciers: *continental glaciers* and *alpine glaciers*. Alpine glaciers are also called *mountain glaciers*.

- Continental glaciers are glaciers that spread out over a large area of land. They were created mostly during the Ice Ages a long time ago. There are still some continental glaciers in Greenland and Antarctica. They often flow downwards into the sea and break up. The broken parts that float in the sea are called icebergs.
- Alpine glaciers form in mountain areas. They are smaller than continental glaciers. Alpine glaciers usually flow until they reach a point where the temperature is warm enough that the ice melts completely during the summer.

Effects

Glaciers are very important. They have a large effect on the environment. They do this because they are very large and heavy. When they move, they erode mountains and land. Also, since they froze long ago, snow crystals and air bubbles inside are kept in good condition. These can provide a large amount of information for scientists. Recently glaciers have been melting more than they did in the past. Many scientists think this is happening because global warming is changing the climate.

Appearance

Glaciers are blue in color. This is because water is very good at absorbing light. Only the strongest light, with the most energy, is able to escape. Blue is the color of light that has the most energy. Because of this, blue is the only color of light that can escape without being absorbed. The thicker the glacier is, the more blue it appears.

The front of Perito Moreno Glacier, Patagonia, Argentina.

GLACIAL LANDFORM

Glacial landforms are landforms created by the action of glaciers. Most of today's glacial landforms were created by the movement of large ice sheets during the Quaternary glaciations. Some areas, like Fennoscandia and the southern Andes, have extensive occurrences of glacial landforms; other areas, such as the Sahara, display very old fossil glacial landforms.

Antique postcard shows rocks scarred by glacial erosion.

Yosemite Valley from an airplane, showing the U-shape.

Glacially-plucked granitic bedrock near Mariehamn, Åland Islands.

Erosional Landforms

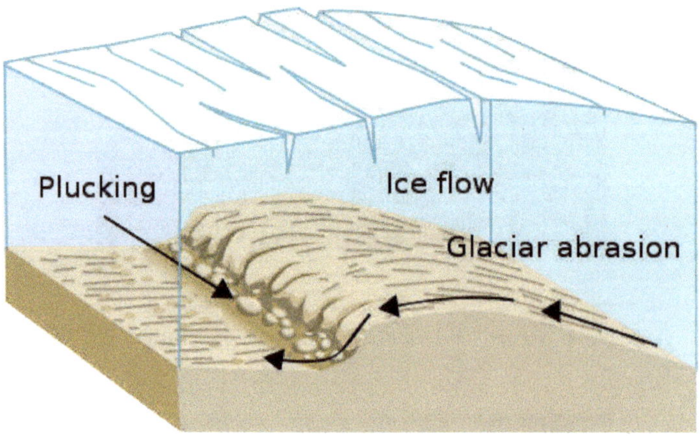

Erosional landforms.

As the glaciers expanded, due to their accumulating weight of snow and ice, they crush and abrade scoured surface rocks and bedrock. The resulting erosional landforms include striations, cirques, glacial horns, arêtes, trim lines, U-shaped valleys, roches moutonnées, overdeepenings and hanging valleys.

- Cirque: Starting location for mountain glaciers
- Cirque stairway: a sequence of cirques
- U-shaped valley: U-shaped valleys are created by mountain glaciers. When filled with ocean water so as to create an inlet, these valleys are called fjords.
- Arête: spiky high land between two glaciers, if the glacial action erodes through, a *spillway* (or col) forms.

Depositional Landforms

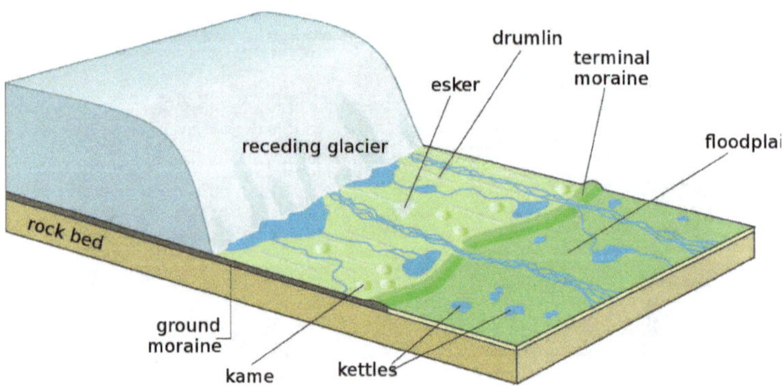

Depositional landforms.

Later, when the glaciers retreated leaving behind their freight of crushed rock and sand (glacial drift), they created characteristic depositional landforms. Examples include glacial moraines, eskers, and kames. Drumlins and ribbed moraines are also landforms left behind by retreating glaciers. The stone walls of New England contain many glacial erratics, rocks that were dragged by a glacier many miles from their bedrock origin.

- Esker: Built up bed of a subglacial stream.
- Kame: Irregularly shaped mound.
- Moraine: Feature can be terminal (at the end of a glacier), lateral (along the sides of a glacier), or medial (formed by the emerger of lateral moraines from contributary glaciers).
- Outwash fan: Braided stream flowing from the front end of a glacier.

Glacial Lakes and Ponds

Lakes and ponds may also be caused by glacial movement. Kettle lakes form when a retreating glacier leaves behind an underground or surface chunk of ice that later melts to form a depression containing water. Moraine-dammed lakes occur when glacial debris dam a stream (or snow runoff). Jackson Lake and Jenny Lake in Grand Teton National Park are examples of moraine-dammed lakes, though Jackson Lake is enhanced by a man-made dam.

- Kettle lake: Depression, formed by a block of ice separated from the main glacier, in which the lake forms.
- Tarn: A lake formed in a cirque by overdeepening.
- Paternoster lake: A series of lakes in a glacial valley, formed when a stream is dammed by successive terminal moraines left by an advancing or retreating glacier.
- Glacial Lake: A lake that formed between the front of a glacier and the last terminal moraine.

Ice Features

Apart from the landforms left behind by glaciers, glaciers themselves may be striking features of the terrain, particularly in the polar regions of the earth. Notable examples include valley glaciers where glacial flow is restricted by the valley walls, crevasses in the upper section of glacial ice, and icefalls—the ice equivalent of waterfalls.

REFERENCES

1. Dreimanis, A., 1988. Tills: Their genetic terminology and classification, p. 17-83. In R. P. Goldthwait and C. L. Matsch, éd., Genetic classification of glacigenic deposits. A. A. Balkema, Rotterdam

2. Meer, J.J.M. van der, Menzies, J. and Rose, J. 2003. Subglacial till: The deforming glacier bed. Quaternary Science Reviews 22, p. 1659-1685.
3. Hambrey, Michael; Alean, Jürg (2004). Glaciers (2nd ed. ed.). Cambridge University Press. ISBN 0-521-82808-2. OCLC 54371738.
4. Benn, Douglas I.; Evans, David J.A. (1999). Glaciers and glaciation. Arnold. ISBN 0470236515. OCLC 38329570.
5. Bennett, M.R.; Glasser N.F. (1996). Glacial geology: ice sheets and landforms. John Wiley & Sons. ISBN 0471963445. OCLC 37536152 33359888 37536152.

CHAPTER 7

Alluvium

Chapter Contents

Alluvium (from the Latin, *alluvius*, from *alluere*, "to wash against") is loose, unconsolidated (not cemented together into a solid rock) soil or sediments, which has been eroded, reshaped by water in some form, and redeposited in a non-marine setting. Alluvium is typically made up of a variety of materials, including fine particles of silt and clay and larger particles of sand and gravel. When this loose alluvial material is deposited or cemented into a lithological unit, or lithified, it is called an alluvial deposit.

INTRODUCTION

The term "alluvium" is not typically used in situations where the formation of the sediment can clearly be attributed to another geologic process that is well described. This includes (but is not limited to): lake sediments (lacustrine), river sediments (fluvial), or glacially-derived sediments (glacial till). Sediments that are formed or deposited in a perennial stream or river are typically not referred to as alluvial. it is also called alluvial soil

Age

Most alluvium is geologically very young (Quaternary in age), and is often referred to as "cover" because these sediments obscure the underlying bedrock. Most sedimentary material that fills a basin ("basin fill") that is not lithified is typically lumped together as "alluvial". Alluvium of Pliocene age occurs, for example, in parts of Idaho. Alluvium of late Miocene age occurs, for example, in the valley of the San Joaquin River, California.

Ores

Alluvium can contain valuable ores such as gold and platinum and a wide variety of gemstones. Such a concentration of valuable ores is termed a placer deposit.

Section of alluvium at the Blue Ribbon Mine in Alaska.

Alluvium deposits in the Gamtoos Valley in South Africa.

An alluvial plain in Red Rock Canyon State Park (California)

Alluvial river deposits in the Amazon Basin, near Autazes, AM, Brazil. The seasonal deposits are extremely fertile, and crucial to subsistence farming in the Amazon Basin along the river banks.

AEOLIAN PROCESSES

Aeolian processes, also spelled eolian or **æolian**, pertain to wind activity in the study of geology and weather and specifically to the wind›s ability to shape the surface of the Earth (or other planets). Winds may erode, transport, and deposit materials and are effective agents in regions with sparse vegetation, a lack of soil moisture and a large supply of unconsolidated sediments. Although water is a much more powerful eroding force than wind, aeolian processes are important in arid environments such as deserts.

The term is derived from the name of the Greek god Aeolus, the keeper of the winds.

Wind erosion of soil at the foot of Chimborazo, Ecuador.

Rock carved by drifting sand below Fortification Rock in Arizona (Photo by Timothy H. O'Sullivan, USGS, 1871).

Wind Erosion

Wind erodes the Earth's surface by deflation (the removal of loose, fine-grained particles by the turbulent action of the wind) and by abrasion (the wearing down of surfaces by the grinding action and sandblasting of windborne particles).

A rock sculpted by wind erosion in the Altiplano region of Bolivia.

Sand blowing off a crest in the Kelso Dunes of the Mojave Desert, California.

Wind-carved alcove in the Navajo Sandstone near Moab, Utah.

Regions which experience intense and sustained erosion are called deflation zones. Most aeolian deflation zones are composed of desert pavement, a sheet-like surface of rock fragments that remains after wind and water have removed the fine particles. Almost half of Earth's desert surfaces are stony deflation zones. The rock mantle in desert pavements protects the underlying material from deflation.

A dark, shiny stain, called desert varnish or rock varnish, is often found on the surfaces of some desert rocks that have been exposed at the surface for a long period of time. Manganese, iron oxides, hydroxides, and clay minerals form most varnishes and provide the shine.

Deflation basins, called blowouts, are hollows formed by the removal of particles by wind. Blowouts are generally small, but may be up to several kilometers in diameter.

Wind-driven grains abrade landforms. In parts of Antarctica wind-blown snowflakes that are technically sediments have also caused abrasion of exposed rocks. Grinding by particles carried in the wind creates grooves or small depressions. Ventifacts are rocks which have been cut, and sometimes polished, by the abrasive action of wind.

Sculpted landforms, called yardangs, are up to tens of meters high and kilometers long and are forms that have been streamlined by desert winds. The famous Great Sphinx of Giza in Egypt may be a modified yardang.

Transport

Particles are transported by winds through suspension, saltation (skipping or bouncing) and creeping (rolling or sliding) along the ground.

Small particles may be held in the atmosphere in suspension. Upward currents of air support the weight of suspended particles and hold them indefinitely in the surrounding air. Typical winds near Earth's surface suspend particles less than 0.2 millimeters in diameter and scatter them aloft as dust or haze.

Dust storm approaching Spearman, Texas April 14, 1935.

Dust storm in Amarillo, Texas. FSA photo by Arthur Rothstein (1936)

A massive sand storm cloud is about to envelop a military camp as it rolls over Al Asad, Iraq, just before nightfall on April 27, 2005.

Saltation is downwind movement of particles in a series of jumps or skips. Saltation normally lifts sand-size particles no more than one centimeter above the ground and proceeds at one-half to one-third the speed of the wind. A saltating grain may hit other grains that to continue the saltation. The grain may also hit larger grains that are too heavy to hop, but that slowly creep forward as they are pushed by saltating grains. Surface creep accounts for as much as 25 percent of grain movement in a desert.

Aeolian turbidity currents are better known as dust storms. Air over deserts is cooled significantly when rain passes through it. This cooler and denser air sinks toward the desert surface. When it reaches the ground, the air is deflected forward and sweeps up surface debris in its turbulence as a dust storm.

Crops, people, villages, and possibly even climates are affected by dust storms. Some dust storms are intercontinental, a few may circle the globe, and occasionally they may engulf entire planets. When the Mariner 9 spacecraft entered its orbit around Mars in 1971, a dust storm lasting one month covered the entire planet, thus delaying the task of photo-mapping the planet's surface.

Most of the dust carried by dust storms is in the form of silt-size particles. Deposits of this windblown silt are known as loess. The thickest known deposit of loess, 335 meters, is on the Loess Plateau in China. This very same Asian dust is blown for thousands of miles, forming deep beds in places as far away as Hawaii. In Europe and in the Americas, accumulations of loess are generally from 20 to 30 meters thick.

Aeolian transport from deserts plays an important role in ecosystems globally, e.g. by transport of minerals from the Sahara to Amazonia. Saharan dust is also responsible for forming red clay soils in southern Europe. Aeolian processes are affected by human activity, such as the use of 4x4 vehicles.

Small whirlwinds, called dust devils, are common in arid lands and are thought to be related to very intense local heating of the air that results in instabilities of the air mass. Dust devils may be as much as one kilometer high.

Deposition

Wind-deposited materials hold clues to past as well as to present wind directions and intensities. These features help us understand the present climate and the forces that molded it. Wind-deposited sand bodies occur as sand sheets, ripples, and dunes.

Sand sheets are flat, gently undulating sandy plots of sand surfaced by grains that may be too large for saltation. They form approximately 40 percent of aeolian depositional surfaces. The Selima Sand Sheet in the eastern Sahara Desert, which occupies 60,000 square kilometers in southern Egypt and northern Sudan, is one of the Earth's largest sand sheets. The Selima is absolutely flat in a few places; in others, active dunes move over its surface.

Cross-bedding of sandstone near Mount Carmel road, Zion National Park, indicating wind action and sand dune formation prior to formation of rock (NPS photo by George A. Grant, 1929)

Mesquite Flat Dunes in Death Valley looking toward the Cottonwood Mountains from the north west arm of Star Dune (2003).

Holocene eolianite deposit on Long Island, The Bahamas. This unit is formed of wind-blown carbonate grains. (2007).

Wind blowing on a sand surface ripples the surface into crests and troughs whose long axes are perpendicular to the wind direction. The average length of jumps during saltation corresponds to the wavelength, or distance between adjacent crests, of the ripples. In ripples, the coarsest materials collect at the crests causing inverse grading. This distinguishes small ripples from dunes, where the coarsest materials are generally in the troughs. This is also a distinguishing feature between water laid ripples and aeolian ripples.

Wind-blown sand moves up the gentle upwind side of the dune by saltation or creep. Sand accumulates at the brink, the top of the slipface. When the buildup of sand at the brink exceeds the angle of repose, a small avalanche of grains slides down the slipface. Grain by grain, the dune moves downwind.

Accumulations of sediment blown by the wind into a mound or ridge, dunes have gentle upwind slopes on the windward side. The downwind portion of the dune, the lee slope, is commonly a steep avalanche slope referred to as a slipface. Dunes may have more than one slipface. The minimum height of a slipface is about 30 centimeters.

Some of the most significant experimental measurements on aeolian sand movement were performed by Ralph Alger Bagnold, a British engineer who worked in Egypt prior to World War II. Bagnold investigated the physics of particles moving through the atmosphere and deposited by wind. He recognized two basic dune types, the crescentic dune, which he called "barchan," and the linear dune, which he called longitudinal or "seif" (Arabic for "sword").

A 2011 study published in *Catena* examined the effect of vegetation on aeolian dust accumulation in the semiarid steppe of northern China. Using a series of trays with different vegetation coverage and a control model with none, the authors found that an increase in vegetation coverage improves the efficiency of dust accumulation and adds more nutrients to the environment, particularly organic carbon. Two critical point were revealed by their data: 1. the efficiency of trapping dust increases slowly above 15% coverage, and decreases rapidly below 15% coverage. 2. at around 55%-75% coverage, dust accumulation reaches a maximum capacity.

A three year quantitative study on the effects of vegetation removal on wind erosion found that the removal of grasses in an aeolian environment increased the rate of soil deposition. In the same study, a relationship was shown between decreasing plant density with decreasing soil nutrients. Similarly, horizontal soil flux across the test site was shown to increase with increasing vegetation removal.

A 1998 study published in Earth Surfaces Processes and Landforms investigated the relationship between vegetative cover on sand surfaces with the rate of sand transport. It was found that sand flux decreased exponentially with vegetation cover. This was done by measuring plots of land with varying degrees of vegetation against rates of sand transport. The authors contend that this relationship can be utilized to manipulate rates of sediment flux by introducing vegetation in an area or to quantify human impact by recognizing vegetation loss's affect on sandy landscapes.

AEOLIAN LANDFORM

Aeolian landforms are features of the Earth's surface produced by either the erosive or constructive action of the wind. This process is not unique to the Earth, and it has been observed and studied on other planets, including Mars.

Terminology

The word derives from Æolus, the Greek god of the winds, and the son of Hellen and the nymph Orseis, and a brother of Dorus, Xuthus and Amphictyon.

Mechanism

Sand blowing off a crest in the Kelso Dunes of the Mojave Desert, California.

In aeolian processes, wind transports and deposits particles of sediment. Aeolian features form in areas where wind is the primary source of erosion. The particles deposited are of sand, silt and clay size. The particles are entrained in by one of four processes. Creep occurs when a particle rolls or slides across the surface. Lift occurs when a particle rises off the surface due to the Bernoulli effect. If the airflow is turbulent, larger particles are transported by a process known as saltation. Finally, impact transport occurs which one particle strikes another causing the second particle to move.

Erosional Landforms

Wind eroded landforms are rarely preserved on the surface of the Earth except in arid regions. Elsewhere, moving water erases the aeolian landforms. There are several types of landforms associated with erosion: lag deposits, ventifacts, yardangs and pans. Large basins are complex and there is often one or more non-aeolian process at work, including tectonics, glacial and alluvial forces.

ALLUVIAL DESERT

A desert pavement, also called reg (in the western Sahara), serir (eastern Sahara), gibber (in Australia), or saï (central Asia) is a desert surface covered with closely packed, interlocking angular or rounded rock fragments of pebble and cobble size. They typically top alluvial fans. Desert varnish collects on the exposed surface rocks over time.

Geologists debate the mechanics of pavement formation and their age.

Desert pavement showing desert varnish on the pebbles; Gibber plains of central Australia.

Desert pavement showing wind scour on the cobbles; Mojave Desert of southern California.

Formation

Several theories have been proposed for the formation of desert pavements. A common theory suggests they form through the gradual removal of sand, dust and other fine-grained material by the wind and intermittent rain, leaving the larger fragments behind. The larger fragments are shaken into place through the forces of rain, running water, wind, gravity, creep, thermal expansion and contraction, wetting and drying, frost heaving, animal traffic, and the Earth's constant microseismic vibrations. The removal of small particles by wind does not continue indefinitely, because once the pavement forms, it acts as a barrier to resist further erosion. The small particles collect underneath the pavement surface, forming a vesicular A soil horizon (designated "Av").

A second theory supposes that desert pavements form from the shrink/swell properties of the clay underneath the pavement; when precipitation is absorbed by clay it causes it to expand, and when it dries it cracks along planes of weakness. Over time, this geomorphic action transports small pebbles to the surface, where they stay through lack of precipitation that would otherwise destroy the pavement by transport of the clasts or excessive vegetative growth.

A newer theory of pavement formation comes from studies of places such as Cima Dome, in the Mojave Desert of California, by Stephen Wells and his coworkers. At Cima Dome, geologically recent lava flows are covered by younger soil layers, with desert pavement on top of them, made of rubble from the same lava. The soil has been built up, not blown away, yet the stones remain on top. There are no stones in the soil, not even gravel.

Researchers can determine how many years a stone has been exposed on the ground. Wells used a method based on cosmogenic helium-3, which forms by cosmic ray bombardment at the ground surface. Helium-3 is retained inside grains of olivine and pyroxene in the lava flows, building up with exposure time. The helium-3 dates show that the lava stones in the desert pavement at Cima Dome have all been at the surface the same amount of time as the solid lava flows right next to them. He wrote in a July 1995 article in *Geology,* that he concluded, "stone pavements are born at the surface." While the stones remain on the surface due to heave, deposition of windblown dust must build up the soil beneath that pavement.

For the geologist, this discovery means that some desert pavements preserve a long history of dust deposition beneath them. The dust is a record of ancient climate, just as it is on the deep sea floor and in the world's ice caps.

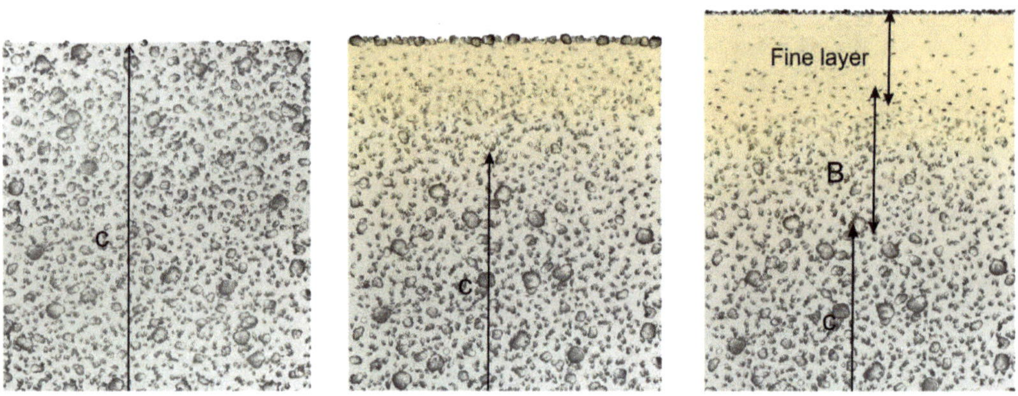

Desert pavement evolution.

Desert pavement surfaces are often coated with desert varnish, a dark brown, sometimes shiny coating that contains clay minerals. In the USA a famous example can be found on Newspaper Rock in southeastern Utah. Desert varnish is a thin coating (patina) of manganese, iron and clays on the surface of sun-baked boulders. It is formed by colonies of bacteria living on the rock surface for thousands of years. Desert varnish is also prevalent in the Mojave desert and Great Basin geomorphic province.

Local Names

Stony deserts may be known by different names according to the region. Examples include:

Gibbers: Covering extensive areas in Australia such as parts of the Tirari-Sturt stony desert ecoregion are desert pavements called *Gibber Plains* after the pebbles or *gibbers*. *Gibber* is also used to describe ecological communities, such as *Gibber Chenopod Shrublands* or *Gibber Transition Shrublands*.

Reg: In North Africa, a vast stony desert plain is known as *reg*. This term is in contrast with *erg*, which refers to a sandy desert area.

The desert known as Reg de l'Adrar in Mauritania.

ALLUVIAL FAN

An alluvial fan is a fan- or cone-shaped deposit of sediment crossed and built up by streams. If a fan is built up by debris flows it is properly called a debris cone or colluvial fan. These flows come from a single point source at the apex of the fan, and over time move to occupy many positions on the fan surface. Fans are typically found where a canyon draining from mountainous terrain emerges out onto a flatter plain, and especially along fault-bounded mountain fronts.

A convergence of neighboring alluvial fans into a single apron of deposits against a slope is called a bajada, or compound alluvial fan.

A vast (60 km long) alluvial fan blossoms across the desolate landscape between the Kunlun and Altun mountain ranges that form the southern border of the Taklamakan Desert in Xinjiang. The left side is the active part of the fan, and appears blue from water flowing in the many small streams.

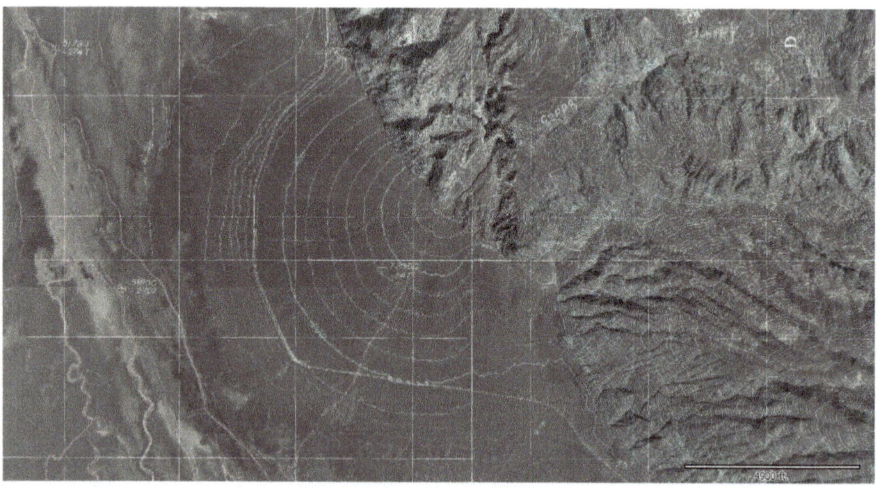

Alluvial fan in Death Valley.

Alluvial fan in the French Pyrenees.

Alluvial fan above Lake Louise, Alberta, Canada.

Formation

As a stream's gradient decreases, it drops coarse-grained material. This reduces the capacity of the channel and forces it to change direction and gradually build up a slightly mounded or shallow conical fan shape. The deposits are usually poorly sorted. This fan shape can also be explained with a thermodynamic justification: the system of sediment introduced at the apex of the fan will tend to a state which minimizes the sum of the transport energy involved in moving the sediment and the gravitational potential of material in the fan. There will be iso-transport energy lines forming concentric arcs about the discharge point at the apex of the fan. Thus the material will tend to be deposited equally about these lines, forming the characteristic fan shape.

In Arid Climates

Alluvial fans are often found in desert areas subject to periodic flash floods from nearby thunderstorms in local hills. The typical watercourse in an arid climate has a large, funnel-shaped basin at the top, leading to a narrow defile, which opens out into an alluvial fan at the bottom. Multiple braided streams are usually present and active during water flows.

Phreatophytes are plants that are often concentrated at the base of alluvial fans. They have long tap roots 30 to 50 feet (9.1 to 15.2 m) to reach water that has seeped through the fan and hit an impermeable layer, sometimes collecting in springs and seeps. These stands of shrubs cling to the soil at their bases and often form islands of habitat for many animals as the wind blows the sand around the bushes away.

In Humid Climates

Alluvial fans also develop in wetter climates. In Nepal the Koshi River has built a megafan covering some 15,000 km^2 (5,800 sq mi) below its exit from Himalayan foothills onto the nearly level plains where the river traverses into India before joining the Ganges. Along the upper Koshi tributaries, tectonic forces elevate the Himalayas several millimeters annually. Uplift is approximately in equilibrium with erosion, so the river annually carries some 100 million cubic meters (3.5 billion cu ft) of sediment as it exits the mountains. Deposition of this magnitude over millions of years is more than sufficient to account for the megafan.

All along the interface between the Indo-Gangetic Plain and the Himalaya in India, Pakistan, Nepal and Bhutan the outermost, lowest Siwalik foothills are built of poorly consolidated sedimentary rocks that have eroded into a wide, continuous alluvial apron called Bhabar in Hindi and Nepali. Despite overpopulation on the plains, this bhabar zone is highly malarial and has remained largely uninhabited.

In North America, streams flowing into California's Central Valley have deposited smaller but still extensive alluvial fans. Such as that of the Kings River flowing out of the Sierra Nevada creates a low divide, turning the south end of the San Joaquin Valley into an endorheic basin without a connection to the ocean.

Flood Hazards

Alluvial fans are subject to flooding and can be even more dangerous than the upstream canyons that feed them. Their slightly convex perpendicular surfaces cause water to spread widely until there is no zone of refuge. If the gradient is steep, active transport of materials down the fan creates a moving substrate that is inhospitable to travel on foot or wheels. But as the gradient diminishes downslope, water comes down from above faster than it can flow away downstream, and may pond to hazardous depths.

In the case of the Koshi River, the huge sediment load and megafan's slightly convex transverse surface conspire against engineering efforts to contain peak flows inside manmade embankments. In August 2008 high monsoon flows breached the embankment, diverting most of the river into an unprotected ancient channel and across surrounding lands with high population density. Over a million people were rendered homeless, about a thousand lost their lives and thousands of hectares of crops were destroyed. The Koshi is known as the Sorrow of Bihar for contributing disproportionately to India's death tolls in flooding, which exceed those of all countries except Bangladesh.

In the Solar System

Alluvial fans are also found on Mars descending from some crater rims over their flatter floors. Observations of fans in Gale crater made by satellites from orbit have now been confirmed by the discovery of fluvial sediments by the Curiosity rover.

Alluvial fans have been observed by the Cassini-Huygens mission on Titan using the Cassini orbiter's synthetic aperture radar (SAR) instrument. These fans are more common in the drier mid-latitudes at the end of methane/ethane rivers where it is thought that frequent wetting and drying occur due to precipitation, much like arid fans on Earth. Radar imaging suggests that fan material is most likely composed of round grains of water ice or solid organic compounds about two centimeters in diameter.

Gallery

Ground-level view of a large alluvial fan in Death Valley.

Small alluvial fan, Death Valley.

Alluvial Fan in southern Iran. Image from NASA's Terra satellite.

Alluvial fan in Death Valley, California, seen from source area of Black Mountains at Dante's View.

ALLUVIAL PLAIN

An alluvial plain is a largely flat landform created by the deposition of sediment over a long period of time by one or more rivers coming from highland regions, from which alluvial soil forms. A floodplain is part of the process, being the smaller area over which the rivers flood at a particular period of time, whereas the alluvial plain is the larger area representing the region over which the floodplains have shifted over geological time.

As the highlands erode due to weathering and water flow, the sediment from the hills is transported to the lower plain. Various creeks will carry the water further to a river, lake, bay, or ocean. As the sediments are deposited during flood conditions in the floodplain of a creek, the elevation of the floodplain will be raised. As this reduces the channel floodwater capacity, the creek will, over time, seek new, lower paths, forming a meander (a curving sinuous path). The leftover higher locations, typically natural levees at the margins of the flood channel, will themselves be eroded by lateral stream erosion and from local rainfall and possibly wind transport if the climate is arid and does not support soil-holding grasses. These processes, over geologic time, will form the plain, a region with little relief (local changes in elevation), yet with a constant but small slope.

The *Glossary of Landform and Geologic Terms*, maintained by the United States' National Cooperative Soil Survey, defines an "alluvial plain" as "a large assemblage of fluvial landforms (braided streams, terraces, etc.,) that form low gradient, regional ramps along the flanks of mountains and extend great distances from their sources (e.g., High Plains of North America)" Use of "alluvial plain" as a general, informal term for a broad flood plain or a low-gradient delta is explicitly discouraged. The NCSS glossary instead suggests "flood plain".

Floodplain (centre) within the alluvial plain of the Waimakariri River, New Zealand (part of the Canterbury Plains).

A small, incised alluvial plain from Red Rock Canyon State Park (California).

Examples

- Mississippi Alluvial Plain
- North China Plain
- Oxnard Plain in Ventura County, California
- Po Valley in Italy
- Punjab alluvial plain
- Indo-Gangetic Plain
- Laguna de Santa Rosa
- Canterbury Plains, Southland Plains, and Waikato Plains in New Zealand
- List of alluvial sites in Switzerland
- Mesopotamia in Iraq and Kuwait
- Mekong Delta, administered as part of Ho Chi Minh City
- Rhine–Meuse–Scheldt delta in the southern Netherlands

REFERENCES

1. Glossary of Geological Terms. Geotech.org. Retrieved on 2012-02-12.
2. Geology Dictionary – Alluvial, Aquiclude, Arkose. Geology.Com. Retrieved on 2012-02-12.
3. Ames, Dan (1998), "Formation of the Soils" (PDF), Soil Survey of Jerome County and Part of Twin Falls County, Idaho, Natural Resources Conservation

Service, USDA, p. 238

4. Huber, N. King (1981). Amount and Timing of Late Cenozoic Uplift and Tilt of the Central Sierra Nevada, California—Evidence from the Upper San Joaquin River Basin (USGS Professional Paper 1197) (PDF). Washington D.C.: USGS. p. 13.
5. National Geographic Almanac of Geography, 2005, page 166, ISBN 0-7922-3877-X.
6. Hsui, Albert T. (2001). "Geology of Mars: Aeolian". Retrieved 2012-09-30.
7. Kurtz, Andrew C; Derry, Louis A; Chadwick, Oliver A (2001). "Accretion of Asian dust to Hawaiian soils: isotopic, elemental, and mineral mass balances" (PDF). Geochimica et Cosmochimica Acta. 65 (12): 1971–1983. doi:10.1016/S0016-7037(01)00575-0. ISSN 0016-7037. Retrieved January 14, 2016.
8. Koren, Ilan; Kaufman, Yoram J; Washington, Richard; Todd, Martin C; Rudich, Yinon; Martins, J Vanderlei; Rosenfeld, Daniel (2006). "The Bodélé depression: a single spot in the Sahara that provides most of the mineral dust to the Amazon forest". Environmental Research Letters. 1 (1): 014005. doi:10.1088/1748-9326/1/1/014005. ISSN 1748-9326. Retrieved January 14, 2016.
9. "African dust caused red soil in southern Europe". Spanish Foundation for Science and Technology. November 11, 2010. Retrieved November 2, 2012.
10. Retta, A.; Wagner, L.E.; Tatarko, J. (2014). "Military Vehicle Trafficking Impacts on Vegetation and Soil Bulk Density at Fort Benning, Georgia" (PDF). Transactions of the ASABE. 57 (4): 1043–1055. doi:10.13031/trans.57.10327. ISSN 2151-0032. Retrieved January 14, 2016.
11. Yan, Yuchun; Xu, Xingliang; Xin, Xiaoping; Yang, Guixia; Wang, Xu; Yan, Ruirui; Chen, Baorui (2011-12-01). "Effect of vegetation coverage on aeolian dust accumulation in a semiarid steppe of northern China". CATENA. 87 (3): 351–356. doi:10.1016/j.catena.2011.07.002.
12. Li, Junran; Okin, Gregory S.; Alvarez, Lorelei; Epstein, Howard (2007-08-08). "Quantitative effects of vegetation cover on wind erosion and soil nutrient loss in a desert grassland of southern New Mexico, USA". Biogeochemistry. 85 (3): 317–332. doi:10.1007/s10533-007-9142-y. ISSN 0168-2563.
13. Lancaster, Nicholas; Baas, Andy (1998-01-01). "Influence of vegetation cover on sand transport by wind: field studies at Owens Lake, California". Earth Surface Processes and Landforms. 23 (1): 69–82. doi:10.1002/(SICI)1096-9837(199801)23:13.0.CO;2-G. ISSN 1096-9837.
14. L. Archie, Ms. Deithra. "Aeolian Processes and Landforms" (PDF). New Mexico State University. p. 2. Retrieved 2008-08-06.
15. "Aeolian landform". The Canadian Encyclopedia. Retrieved 2008-08-06.
16. American Geological Institute. Dictionary of Geological Terms. New York: Dolphin Books, 1962.
17. To clarify, solids are sorted as usual, with coarse sediment dropped out first -- but the sorting of an individual flood event is then "jumbled" by the next

flood, leaving the overall fan sediment package poorly sorted.

18. National Aeronautics and Space Administration. "Geomorphology from Space; Fluvial Landforms, Chapter 4: Plate F-19". Retrieved April 18, 2009.
19. Cazanacli, Dan; Paola, Chris; Parker, Gary (2002). "Experimental Steep, Braided Flow: Application to Flooding Risk on Fans". Journal of Hydraulic Engineering. 128 (3): 322. doi:10.1061/(ASCE)0733-9429(2002)128:3(322).
20. Committee on Alluvial Fan Flooding, Water Science and Technology Board, Commission on Geosciences, Environment, and Resources, National Research Council. (1996). Alluvial fan flooding. Washington, D.C.: National Academy Press. ISBN 0-309-05542-3.
21. Kraal, Erin R.; Asphaug, Erik; Moore, Jeffery M.; Howard, Alan; Bredt, Adam (March 2008). "Catalogue of large alluvial fans in martian impact craters". Icarus. 194 (1): 101–110. doi:10.1016/j.icarus.2007.09.028. ISSN 0019-1035. Retrieved January 21, 2016.
22. Harwood, William; Wall, Mike (September 27, 2012). "Mars rover Curiosity finds ancient stream bed". CBS News. Retrieved January 21, 2016.
23. J. Radebaugh; et al. (2013). "Alluvial Fans on Titan Reveal Materials, Processes and Regional Conditions" (PDF). 44th Lunar and Planetary Science Conference. Retrieved January 21, 2016.
24. "Glossary of Landform and Geologic Terms" (PDF). National Soil Survey Handbook—Part 629. National Cooperative Soil Survey. April 2013. Retrieved 17 August 2016.
25. Mississippi River alluvial plain
26. Punjab Plain, Encyclopædia Britannica

CHAPTER 8

Stratigraphy

Chapter Contents

- **Introduction**
- **Historical Development**
- **Lithostratigraphy**
- **Biostratigraphy**
- **Chronostratigraphy**
- **Magnetostratigraphy**
- **Chemostratigraphy**
- **Tectonostratigraphy**
- **Cyclostratigraphy**
- **Archaeological Stratigraphy**

Stratigraphy is the study of rock layers (strata) deposited in the earth. It is one of the most challenging of geologic subdisciplines, comparable to an exacting form of detective work, yet it is also one of the most important branches of study in the geologic sciences. Earth's history, quite literally, is written on the strata of its rocks, and from observing these layers, geologists have been able to form an idea of the various phases in that long history. Naturally, information is more readily discernible about the more recent phases, though even in studying these phases, it is possible to be misled by gaps in the rock record, known as unconformities.

INTRODUCTION

Stratigraphy is a branch of geology which studies rock layers (strata) and layering (stratification). It is primarily used in the study of sedimentary and layered volcanic rocks. Stratigraphy has two related subfields: lithologic stratigraphy or lithostratigraphy, and biologic stratigraphy or biostratigraphy.

The Permian through Jurassic strata of the Colorado Plateau area of southeastern Utah demonstrate the principles of stratigraphy.

HISTORICAL DEVELOPMENT

Engraving from William Smith's monograph on identifying strata based on fossils.

Nicholas Steno established the theoretical basis for stratigraphy when he introduced the law of superposition, the principle of original horizontality and the principle of lateral continuity in a 1669 work on the fossilization of organic remains in layers of sediment.

The first practical large-scale application of stratigraphy was by William Smith in the 1790s and early 19th century. Smith, known as the "Father of English geology", created the first geologic map of England and first recognized the significance of strata or rock layering and the importance of fossil markers for correlating strata. Another influential application of stratigraphy in the early 19th century was a study by Georges Cuvier and Alexandre Brongniart of the geology of the region around Paris.

Strata in Cafayate (Argentina)

LITHOSTRATIGRAPHY

Lithostratigraphy is a sub-discipline of stratigraphy, the geological science associated with the study of strata or rock layers. Major focuses include geochronology, comparative geology, and petrology. In general a stratum will be primarily igneous or sedimentary relating to how the rock was formed.

Sedimentary layers are laid down by deposition of sediment associated with weathering processes, decaying organic matters (biogenic) or through chemical precipitation. These layers are distinguishable as having many fossils and are important for the study of biostratigraphy. Igneous layers are either plutonic or volcanic in character depending upon the cooling rate of the rock. These layers are generally devoid of fossils and represent intrusions and volcanic activity that occurred over the geologic history of the area.

There are a number of principles that are used to explain the appearance of stratum. When an igneous rock cuts across a formation of sedimentary rock, then we can say that the igneous intrusion is younger than the sedimentary rock. The principle of superposition states that a sedimentary rock layer in a tectonically undisturbed stratum is younger than the one beneath and older than the one above it. The principle of original horizontality states that the deposition of sediments occurs as essentially horizontal beds.

Types of Lithostratigraphic Units

A lithostratigraphic unit conforms to the law of superposition, which state that in any succession of strata, not disturbed or overturned since deposition, younger rocks lies above older rocks. The principle of lateral continuity states that a set of bed extends and can be traceable over a large area.

Lithostratigraphic units are recognized and defined on the basis of observable rock characteristics. The descriptions of strata based on physical appearance define facies. Lithostratigraphic units are only defined by lithic characteristics, and not by age.

Stratotype: A designated type of unit consisting of accessible rocks that contain clear-cut characteristics which are representative of a particular lithostratigraphic unit.

Lithosome: Masses of rock of essentially uniform character and having interchanging relationships with adjacent masses of different lithology. e.g.: shale lithosome, limestone lithosome.

The fundamental Lithostratigraphic unit is the formation. A formation is a lithologically distinctive stratigraphic unit that is large enough to be mappable and traceable. Formations may be subdivided into members and beds and aggregated with other formations into groups and supergroups.

Stratigraphic Relationship

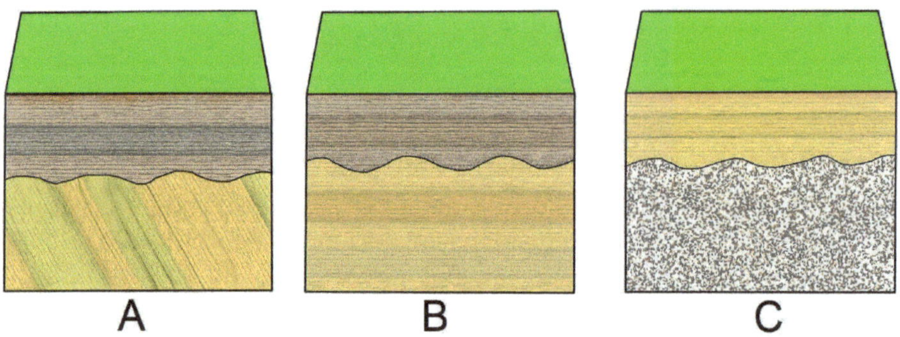

Diagrams showing stratigraphic relations: A: an angular unconformity; B: a disconformity; C: a nonconformity.

Two types of contact: *conformable* and *unconformable*.

- *Conformable*: unbroken deposition, no break or hiatus (break or interruption in the continuity of the geological record). The surface strata resulting is called a *conformity*.

  Two types of contact between conformable strata: *abrupt contacts* (directly separate beds of distinctly different lithology, minor depositional break, called *diastems*) and *gradational contact* (gradual change in deposition, mixing zone).

- *Unconformable*: period of erosion/non-deposition. The surface stratum resulting is called an unconformity.

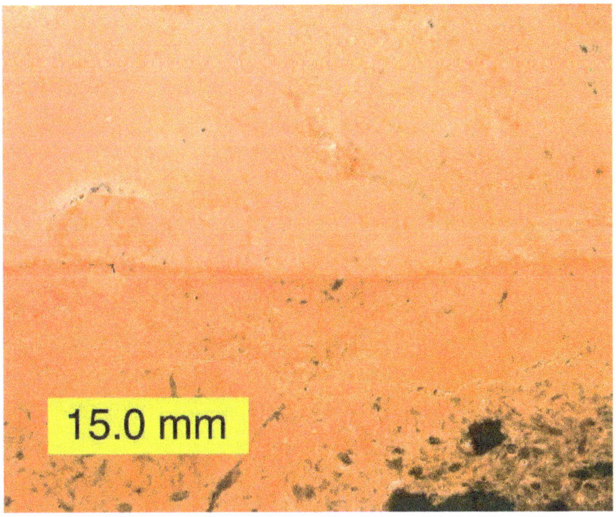

Disconformity with the Lower Cretaceous Edwards Formation overlying a Lower Permian limestone; hiatus is about 165 million years; Texas.

Four types of unconformity:

- *Angular unconformity*: younger sediment lies upon an eroded surface of tilted or folded older rocks. The older rock dips at a different angle from the younger.
- *Disconformity*: the contact between younger and older beds is marked by visible, irregular erosional surfaces. Paleosol might develop right above the disconformity surface because of the non-deposition setting.
- *Paraconformity*: the bedding planes below and above the unconformity are parallel. A time gap is present, as shown by a faunal break, but there is no erosion, just a period of non-deposition.
- *Nonconformity*: relatively young sediments are deposited right above older igneous or metamorphic rocks.

Lithostratigraphic Correlation

To correlate lithostratigraphic units, geologists define facies, and look for key beds or key sequences that can be used as a datum.

- Direct correlation: based on lithology, color, structure, thickness...
- Indirect correlation: electric log correlation (gamma-ray, density, resistivity...)

BIOSTRATIGRAPHY

Biostratigraphy is the branch of stratigraphy which focuses on correlating and assigning relative ages of rock strata by using the fossil assemblages contained within them. Usually the aim is correlation, demonstrating that a particular horizon in one geological section represents the same period of time as another horizon at some other section. The fossils are useful because sediments of the same age can look completely different because of local variations in the sedimentary environment. For example, one section might have been made up of clays and marls while another has more chalky limestones, but if the fossil species recorded are similar, the two sediments are likely to have been laid down at the same time.

Biostratigraphy originated in the early 19th century, where geologists recognized that the correlation of fossil assemblages between rocks of similar type but different age decreased as the difference in age increased. The method was well-established before Charles Darwin explained the mechanism behind it - evolution.

The first reef builder is a worldwide index fossil for the Lower Cambrian.

Ammonites, graptolites, archeocyathids, and trilobites are index fossils that are widely used in biostratigraphy. Microfossils such as acritarchs, chitinozoans, conodonts, dinoflagellate cysts, pollen, spores and foraminiferans are also frequently used. Different fossils work well for sediments of different ages; trilobites, for example, are particularly useful for sediments of Cambrian age. To work well, the fossils used must be widespread geographically, so that they can occur in many different places. They must also be short lived as a species, so that the period of time during which they could be incorporated in the sediment is relatively narrow. The longer lived the species, the poorer the stratigraphic precision, so fossils that evolve rapidly, such as ammonites, are favoured over forms that evolve much more slowly, like nautiloids. Often biostratigraphic correlations are based on a fauna, not an individual species, as this allows greater precision. Further, if only one species is present in a sample, it can mean that (1) the strata were formed in the known fossil range of that organism; (2) that the fossil range of the organism was incompletely known, and the strata extend the known fossil range. For instance, the presence of the fossil *Treptichnus pedum* was used to define the base of the Cambrian period, but it has since been found in older strata.

Fossil assemblages were traditionally used to designate the duration of periods. Since a large change in fauna was required to make early stratigraphers create a new period, most of the periods we recognize today are terminated by a major extinction event or faunal turnover.

Fossils as a Basis for Stratigraphic Subdivision

Concept of Stage

A stage is a major subdivision of strata, each systematically following the other each bearing a unique assemblage of fossils. Therefore, stages can be defined as a group of strata containing the same major fossil assemblages. French palaeontologist Alcide d'Orbigny is credited for the invention of this concept. He named stages after geographic localities with particularly good sections of rock strata that bear the characteristic fossils on which the stages are based.

Concept of Zone

In 1856 German palaeontologist Albert Oppel introduced the concept of zone (also known as biozones or Oppel zone). A zone includes strata characterized by the overlapping range of fossils. They represent the time between the appearance of species chosen at the base of the zone and the appearance of other species chosen at the base of the next succeeding zone. Oppel's zones are named after a particular distinctive fossil species, called an index fossil. Index fossils are one of the species from the assemblage of species that characterize the zone.

The zone is the fundamental biostratigraphic unit. Its thickness range from a few to hundreds of meters, and its extant range from local to worldwide. Biostratigraphic units are divided into six principal kinds of biozones:

- *Taxon range biozone* represent the known stratigraphic and geographic range of occurrence of a single taxon.
- *Concurrent range biozone* include the concurrent, coincident, or overlapping part of the range of two specified taxa.
- *Interval biozone* include the strata between two specific biostratigraphic surfaces. It can be based on lowest or highest occurrences.
- *Lineage biozone* are strata containing species representing a specific segment of an evolutionary lineage.

- *Assemblage biozones* are strata that contain a unique association of three or more taxa.
- *Abundance biozone* are strata in which the abundance of a particular taxon or group of taxa is significantly greater than in the adjacent part of the section.

Index Fossils

Amplexograptus, a graptolite index fossil, from the Ordovician near Caney Springs, Tennessee.

To be useful in stratigraphic correlation index fossils should be:

- Independent of their environment
- Geographically widespread (provincialism/isolation of species should be avoided as much as possible)
- Rapidly evolving
- Abundant (easy to find in the rock record)
- Easy to preserve (Easier in low-energy, non-oxidized environment)
- Easy to identify

Conodont Biostratigraphy

Conodonts are an extinct class of animals whose feeding apparatuses called *teeth* or *elements* are common microfossils found in strata dating from the Stage 10 of the Furongian, the fourth and final series of the Cambrian, to the Rhaetian stage of the Late Triassic. These elements can be used alternatively to or in correlation with other types of fossils (graptolites, trilobites, ammonites, ...) in the subfield of the stratigraphy named biostratigraphy.

Paleozoic Conodonts

Cambrian Conodonts

It is suggested that *Eoconodontus notchpeakensis* can be a marker of the Stage 10 of the Furongian, the fourth and final series of the Cambrian.

In 2006, a working group proposed the first appearance of *Cordylodus andresi*. Currently the first appearance of *E. notchpeakensis* is favored by many authors because it is globally widespread and is independent of facies (known from continental rise to peritidal environments).

The *Eoconodontus notchpeakensis* proposal would also incorporate a non-biostratigraphic marker to correlate the beginning of Stage 10 globally. A carbon isotope excursion (the HERB-event) occurs in the lower part of the E. notchpeakensis range.

Ordovician Conodonts

Early Ordovician

The base of the Tremadocian, the lowest stage of Ordovician, is defined as the first appearance of *Iapetognathus fluctivagus* at the GSSP section in Newfoundland.

No conodont species is associated with the Floian, the second stage of the Ordovician.

Middle Ordovician

The base of the Dapingian, the third stage of the Ordovician, is defined as the first appearance of *Baltoniodus triangularis*.

The base of the Darriwilian, the fourth stage of the Ordovician, lies just above the North Atlantic *Microzarkodina parva* conodont zone. The base also lies in the upper part of the North American *Histiodella altifrons* conodont zone.

The Whiterock Stage refers mainly to the early Middle Ordovician in North America. It is often used in the older literature in a global sense. The Whiterock Stage is given a range from 471.8 (ca. 472) to 462 m.y.a., spanning close to 10 million years. Officially its start is defined by the potentially lowest occurrence of the conodont *Protoprioniodus aranda* or *Baltoniodus triangularis*.

Late Ordovician

No conodont species is associated with the Sandbian, the Katian, nor with the Hirnantian, the fifth, sixth and seventh and final stages of the Ordovician.

Baltoniodus gerdae has been found in the early Sandbian Bromide Formation, in Oklahoma, USA.

Silurian Conodonts

Llandovery

The Llandovery Epoch lasted from 443.8 ± 1.5 to 433.4 ± 2.8 mya, and is subdivided into three stages: the Rhuddanian, lasting until 440.8 million years ago, the Aeronian, lasting to 438.5 million years ago, and the Telychian.

The Telychian (Late Llandovery) of Estonia can be defined by five conodont zones (*Pterospathodus eopennatus ssp. n. 1*, *P. eopennatus ssp. n. 2*, *P. amorphognathoides angulatus*, *P. a. lennarti* and *P. a. lithuanicus*).

Wenlock

The Wenlock, which lasted from 433.4 ± 1.5 to 427.4 ± 2.8 mya, is subdivided into the Sheinwoodian (to 430.5 million years ago) and Homerian ages.

The Sheinwoodian (Wenlock) is defined between the acritarch biozone 5 and the last appearance of *Pterospathodus amorphognathoides*. The global boundary stratotype point is in Hughley Brook in Apedale, U.K.

Ludlow

The Ludlow, lasting from 427.4 ± 1.5 to 423 ± 2.8 mya, comprises the Gorstian stage, lasting until 425.6 million years ago, and the Ludfordian stage.

Devonian Conodonts

Early Devonian

The Early Devonian lasts from 419.2 ± 2.8 to 393.3 ± 2.5 and begins with the Lochkovian stage, which lasts until the Pragian. This spans from 410.8 ± 2.8 to 407.6 ± 2.5, and is followed by the Emsian, which lasts until the Middle Devonian begins, 393.3± 2.7 million years ago.

Middle Devonian

The Middle Devonian comprises two subdivisions, the Eifelian giving way to the Givetian 387.7± 2.7 million years ago.

Late Devonian

Finally, the Late Devonian starts with the Frasnian, 382.7 ± 2.8 to 372.2 ± 2.5, followed by the Famennian subdivision, the beginning and end of which are marked with extinction events. This lasted until the end of the Devonian, 358.9± 2.5 million years ago. During that stage, a biologic event occurred (Upper Kellwasser Extinction of all *Ancyrodella* and *Ozarkodina* and most *Palmatolepis*, *Polygnathus* and *Ancyrognathus*). The Famennian (372.2

± 1.6 mya) is defined by a GSSP Golden Spike located at Coumiac quarry, Montagne Noire, France where there is a biologic abundant occurrence of *Palmatolepis triangularis*.

Carboniferous Conodonts

Mississipian (also known as Lower Carboniferous)

The Tournaisian, the oldest age of the Mississippian contains eight conodont biozones:

- the zone of *Gnathodus pseudosemiglaber* and *Scaliognathus anchoralis*
- the zone of *Gnathodus semiglaber* and *Polygnathus communis*
- the zone of *Dollymae bouckaerti*
- the zone of *Gnathodus typicus* and *Siphonodella isosticha*
- the zone of *Siphonodella quadruplicata* and *Patrognathus andersoni* (upper zone of *Patrognathus andersoni*)
- the lower zone of *Patrognathus andersoni*
- the zone of *Patrognathus variabilis*
- the zone of *Patrognathus crassus*

The Visean, the second age of the Mississippian, contains four conodont biozones:

- the zone of *Lochriea nodosa*
- the zone of *Lochriea mononodosa*
- the zone of *Gnathodus bilineatus*
- the zone of *Gnathodus texanus*

The Serpukhovian, the third or youngest age of the Mississippian, includes four conodont biozones:

- the zone of *Gnathodus postbilineatus*
- the zone of *Gnathodus bollandensis*
- the zone of *Lochriea cruciformis*
- the zone of *Lochriea ziegleri*

Pennsylvanian (also known as Upper Carboniferous)

The Bashkirian, the oldest age of the Pennsylvanian, contains six biozones based on conodont index fossils:

- the zone of *Neognathodus atokaensis*
- the zone of *Declinognathodus marginodosus*
- the zone of *Idiognathodus sinuosus*
- the zone of *Neognathodus askynensis*
- the zone of *Idiognathoides sinuatus*
- the zone of *Declinognathodus noduliferus*

The base of the Moscovian, the second stage in the Pennsylvanian, is close to the first appearances of the conodonts *Declinognathodus donetzianus* and *Idiognathoides postsulcatus*. A proposal is to use the first appearance of the conodont *Diplognathodus ellesmerensis*, but since the species is rare and its evolution relatively unknown, it has not been accepted yet. The Moscovian can biostratigraphically be divided into five conodont biozones:

- the zone of *Neognathodus roundyi* and *Streptognathodus cancellosus*
- the zone of *Neognathodus medexultimus* and *Streptognathodus concinnus*
- the zone of *Streptognathodus dissectus*
- the zone of *Neognathodus uralicus*
- the zone of *Declinognathodus donetzianus*

The top of the Kasimovian, the third stage in the Pennsylvanian, is close to the first appearance of the conodont *Streptognathodus zethus*. The Kasimovian is subdivided into three conodont biozones:

- the zone of *Idiognathodus toretzianus*
- the zone of *Idiognathodus sagittatus*
- the zone of *Streptognathodus excelsus* and *Streptognathodus makhlinae*

The base of the Gzhelian, the youngest age of the Pennsylvanian, is at the first appearance of *Streptognathodus zethus*. The top of the stage (the base of the Permian system) is at the first appearance of *Streptognathodus isolatus* within the *Streptognathus "wabaunsensis"* chronocline. The Gzhelian stage is subdivided into five biozones, based on the conodont genus *Streptognathodus*:

- the zone of *Streptognathodus wabaunsensis* and *Streptognathodus bellus*
- the zone of *Streptognathodus simplex*
- the zone of *Streptognathodus virgilicus*
- the zone of *Streptognathodus vitali*
- the zone of *Streptognathodus simulator*

Permian Conodonts

Cisuralian

The base of the Asselian stage is at the same time the base of the Cisuralian series and the Permian system. It is defined as the place in the stratigraphic record where fossils of *Streptognathodus isolatus* first appear. The top of the Asselian stage (the base of the Sakmarian stage) is at the first appearance of conodont species *Streptognathodus postfusus*. The Asselian contains five conodont biozones:

- the zone of *Streptognathodus barskovi*
- the zone of Streptognathodus postfusus
- the zone of Streptognathodus fusus

- the zone of Streptognathodus constrictus
- the zone of Streptognathodus isolatus

The base of the Sakmarian stage is laid with the first appearance of *Streptognathodus postfusus* in the fossil record. The top of the Sakmarian (the base of the Artinskian) is defined as the place in the stratigraphic record where fossils of *Sweetognathus whitei* and *Mesogondolella bisselli* first appear.

The base of the Artinskian stage is defined as the place in the stratigraphic record where fossils of *Sweetognathus whitei* and *Mesogondolella bisselli* first appear. The top of the Artinskian (the base of the Kungurian) is defined as the place in the stratigraphic record where fossils of *Neostreptognathodus pnevi* and *Neostreptognathodus exculptus* first appear.

The base of the Kungurian stage is defined as the place in the stratigraphic record where fossils of *Neostreptognathodus pnevi* and *Neostreptognathodus exculptus* first appear. The top of the Kungurian (the base of the Roadian and the Guadalupian series) is defined as the place in the stratigraphic record where fossils of *Jinogondolella nanginkensis* first appear. The Kungurian contains three conodont biozones:

- the zone of *Neostreptognathodus sulcoplicatus*
- the zone of Neostreptognathodus prayi
- the zone of Neostreptognathodus pnevi

Guadalupian

The base of the Roadian is defined as the place in the stratigraphic record where fossils of *Jinogondolella nankingensis* first appears. The top of the Roadian (the base of the Wordian stage) is at the first appearance of fossils of *Jinogondolella aserrata*.

The base of the Wordian stage is defined as the place in the stratigraphic record where fossils of *Jinogondolella aserrata* first appear. The top of the Wordian (the base of the Capitanian stage) is defined as the place in the stratigraphic record where *Jinogondolella postserrata* first appears.

The base of the Capitanian stage is defined as the place in the stratigraphic record where fossils of *Jinogondolella postserrata* first appear. The top of the Capitanian (the base of the Wuchiapingian and Lopingian series) is defined as the place in the stratigraphic record where *Clarkina postbitteri postbitteri* first appears. The Capitanian contains three conodont biozones:

- the zone of *Clarkina postbitteri hongshuiensis*
- the zone of *Jinogondolella altudaensis*
- the zone of *Jinogondolella postserrata*

Lopingian

The base of the Wuchiapingian stage is defined as the place in the stratigraphic record where *Clarkina postbitteri postbitteri* first appears. The top of the Wuchiapingian (the base of the Changhsingian) is at the first appearance of conodont species *Clarkina wangi*.

The base of the Changhsingian stage is at the first appearance of *Clarkina wangi*. The top of the Changhsingian (the base of the Induan stage and the Triassic system) is at the first appearance of *Hindeodus parvus*.

Mesozoic Conodonts

Early Triassic

The base of the Induan stage (which is also the base of the Lower Triassic series, the base of the Triassic system and the base of the Mesozoic erathem) is defined as the place in the fossil record where *Hindeodus parvus* first appears.

The base of the Olenekian is at the lowest occurrence of *Neospathodus waageni*. It is defined as ending near the lowest occurrences of *Chiosella timorensis*.

Middle Triassic

The base of the Anisian stage (also the base of the Middle Triassic series) is sometimes laid at the first appearance of *Chiosella timorensis* in the stratigraphic record. The top of the Anisian (the base of the Ladinian) is at the first appearance of *Neogondolella praehungarica*.

The Ladinian is defined by the first appearance of *Budurovignathus praehungaricus*.

Upper Triassic

The top of the Carnian (the base of the Norian) is at the conodont biozones of *Metapolygnathus communisti* or *Metapolygnathus primitius*.

The Norian stage begins at the base of the conodont biozones of *Metapolygnathus communisti* and *Metapolygnathus primitius*. The top of the Norian (the base of the Rhaetian) is close to the first appearance of *Misikella spp.* and *Epigondolella mosheri*.

The Alaunian, also known as "Middle Norian", is a sub-age in the Upper Triassic. It begins with the first appearance of *Cypridodella multidentata*. The stage ends with the first appearance of *Cypridodella bidentata*.

CHRONOSTRATIGRAPHY

Chronostratigraphy is the branch of stratigraphy that studies the age of rock strata in relation to time.

The ultimate aim of chronostratigraphy is to arrange the sequence of deposition and the time of deposition of all rocks within a geological region, and eventually, the entire geologic record of the Earth.

The standard stratigraphic nomenclature is a chronostratigraphic system based on palaeontological intervals of time defined by recognised fossil assemblages (biostratigraphy).

The aim of chronostratigraphy is to give a meaningful age date to these fossil assemblage intervals and interfaces.

Methodology

Chronostratigraphy relies heavily upon isotope geology and geochronology to derive hard dating of known and well defined rock units which contain the specific fossil assemblages defined by the stratigraphic system. As it is practically very difficult to isotopically date most fossils and sedimentary rocks directly, inferences must be made in order to arrive at an age date which reflects the beginning of the interval.

The methodology used is derived from the law of superposition and the principles of cross-cutting relationships.

Because igneous rocks occur at specific intervals in time and are essentially instantaneous on a geologic time scale, and because they contain mineral assemblages which may be dated more accurately and precisely by isotopic methods, the construction of a chronostratigraphic column relies heavily upon intrusive and extrusive igneous rocks.

Metamorphism, often associated with faulting, may also be used to bracket depositional intervals in a chronostratigraphic column. Metamorphic rocks can occasionally be dated, and this may give some limits to the age at which a bed could have been laid down. For example, if a bed containing graptolites overlies crystalline basement at some point, dating the crystalline basement will give a maximum age of that fossil assemblage.

This process requires a considerable degree of effort and checking of field relationships and age dates. For instance, there may be many millions of years between a bed being laid down and an intrusive rock cutting it; the estimate of age must necessarily be between the oldest cross-cutting intrusive rock in the fossil assemblage and the youngest rock upon which the fossil assemblage rests.

Units

Chronostratigraphic units, with examples:

- eonothem – Phanerozoic
- erathem – Paleozoic
- system – Ordovician
- series – Upper Ordovician
- stage – Ashgill

Differences between Chronostratigraphy and Geochronology

It is important not to confuse geochronologic and chronostratigraphic units. Chronostratigraphic units are geological material, so it is correct to say that fossils of the species *Tyrannosaurus rex* have been found in the Upper Cretaceous Series. Geochronological units are periods of time and take the same name as standard stratigraphic units but replacing the terms upper/

lower with late/early. Thus it is also correct to say that *Tyrannosaurus rex* lived during the Late Cretaceous Epoch. Chronostratigraphy is an important branch of stratigraphy because the age correlations derived are crucial to drawing accurate cross sections of the spatial organization of rocks and to preparing accurate paleogeographic reconstructions.

MAGNETOSTRATIGRAPHY

Magnetostratigraphy is a geophysical correlation technique used to date sedimentary and volcanic sequences. The method works by collecting oriented samples at measured intervals throughout the section. The samples are analyzed to determine their *characteristic remanent magnetization* (ChRM), that is, the polarity of Earth's magnetic field at the time a stratum was deposited. This is possible because volcanic flows acquire a thermoremanent magnetization and sediments acquire a depositional remanent magnetization, both of which reflect the direction of the Earth's field at the time of formation.

Technique

When measurable magnetic properties of rocks vary stratigraphically they may be the basis for related but different kinds of stratigraphic units known collectively as *magnetostratigraphic units (magnetozones)*. The magnetic property most useful in stratigraphic work is the change in the direction of the remanent magnetization of the rocks, caused by reversals in the polarity of the Earth's magnetic field. The direction of the remnant magnetic polarity recorded in the stratigraphic sequence can be used as the basis for the subdivision of the sequence into units characterized by their magnetic polarity. Such units are called "magnetostratigraphic polarity units" or chrons.

If the ancient magnetic field was oriented similar to today's field (North Magnetic Pole near the Geographic North Pole) the strata retain a normal polarity. If the data indicate that the North Magnetic Pole was near the Geographic South Pole, the strata exhibit reversed polarity.

Sampling Procedures

Oriented paleomagnetic samples are collected in the field using a rock core drill, or as *hand samples* (chunks broken off the rock face). To average out sampling errors, a minimum of three samples is taken from each sample site. Spacing of the sample sites within a stratigraphic section depends on the rate of deposition and the age of the section. In sedimentary layers, the preferred lithologies are mudstones, claystones, and very fine-grained siltstones because the magnetic grains are finer and more likely to orient with the ambient field during deposition.

Analytical Procedures

Samples are first analyzed in their natural state to obtain their natural remanent magnetization (NRM). The NRM is then stripped away in a stepwise manner using thermal or alternating field demagnetization techniques to reveal the stable magnetic component.

Magnetic orientations of all samples from a site are then compared and their average magnetic polarity is determined with directional statistics, most commonly Fisher statistics or bootstrapping. The statistical significance of each average is evaluated. The latitudes of the Virtual Geomagnetic Poles from those sites determined to be statistically significant are plotted against the stratigraphic level at which they were collected. These data are then abstracted to the standard black and white magnetostratigraphic columns in which black indicates normal polarity and white is reversed polarity.

Correlation and Ages

Because the polarity of a stratum can only be normal or reversed, variations in the rate at which the sediment accumulated can cause the thickness of a given polarity zone to vary from one area to another. This presents the problem of how to correlate zones of like polarities between different stratigraphic sections. To avoid confusion at least one isotopic age needs to be collected from each section. In sediments, this is often obtained from layers of volcanic ash. Failing that, one can tie a polarity to a biostratigraphic event that has been correlated elsewhere with isotopic ages. With the aid of the independent isotopic age or ages, the local magnetostratigraphic column is correlated with the Global Magnetic Polarity Time Scale (GMPTS).

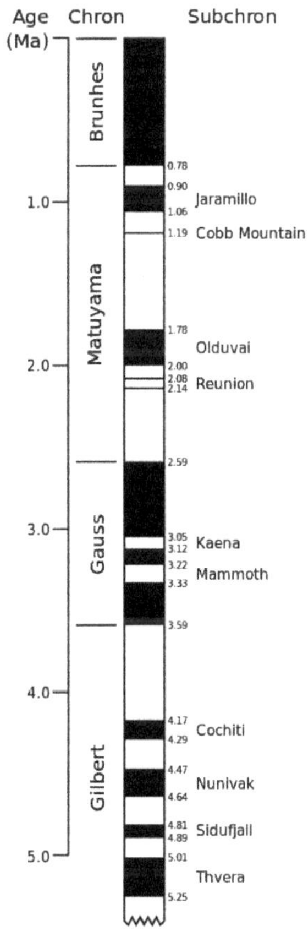

Geomagnetic polarity in late Cenozoic. normal polarity (black) reverse polarity (white).

Because the age of each reversal shown on the GMPTS is relatively well known, the correlation establishes numerous time lines through the stratigraphic section. These ages provide relatively precise dates for features in the rocks such as fossils, changes in sedimentary rock composition, changes in depositional environment, etc. They also constrain the ages of cross-cutting features such as faults, dikes, and unconformities.

Sediment Accumulation Rates

Perhaps the most powerful application of these data is to determine the rate at which the sediment accumulated. This is accomplished by plotting the age of each reversal (in millions of years ago) vs. the stratigraphic level at which the reversal is found (in meters). This provides the rate in meters per million years which is usually rewritten in terms of millimeters per year (which is the same as kilometers per million years).

These data are also used to model basin subsidence rates. Knowing the depth of a hydrocarbon source rock beneath the basin-filling strata allows calculation of the age at which the source rock passed through the generation window and hydrocarbon migration began. Because the ages of cross-cutting trapping structures can usually be determined from magnetostratigraphic data, a comparison of these ages will assist reservoir geologists in their determination of whether or not a play is likely in a given trap.

Changes in sedimentation rate revealed by magnetostratigraphy are often related to either climatic factors or to tectonic developments in nearby or distant mountain ranges. Evidence to strengthen this interpretation can often be found by looking for subtle changes in the composition of the rocks in the section. Changes in sandstone composition are often used for this type of interpretation.

CHEMOSTRATIGRAPHY

Chemostratigraphy, or chemical stratigraphy, is the study of the chemical variations within sedimentary sequences to determine stratigraphic relationships. The field is relatively young, having only come into common usage in the early 1980s, but the basic idea of chemostratigraphy is nearly as old as stratigraphy itself: distinct chemical signatures can be as useful as distinct fossil assemblages or distinct lithographies in establishing stratigraphic relationships between different rock layers.

Types of Chemical Variations

In some stratigraphic sequences, there is clearly a variation in color between different strata. Such color differences often originate from variations in the incorporation of transition metal-containing materials during deposition and lithifaction. Other differences in color can originate from variations in the organic carbon content of the rock. However, until relatively recently, these variations were not commonly investigated because of the great effort and expense involved in chemical analysis.

Recently, the development of new analytical techniques for chemical analysis for igneous petrological applications during the latter half of the 20th century, e.g., the electron microprobe, and the development of normal focus X-ray fluorescence for wellsite oil exploration has improved the availability of bulk chemical analysis techniques to the sedimentary geologist, making analysis of the chemical composition of strata increasingly possible. Concurrently, advances in atomic physics stimulated investigations in stable isotope geochemistry. Most relevant to chemostratigraphy in general was the discovery by Harold Urey and Cesare Emiliani in the early 1950s that the oxygen isotope variability in the calcite shells of foraminifera could be used as a proxy for past ocean temperatures.

Thus, chemostratigraphy generally provides two useful types of information to the larger geological community. First, chemostratigraphy can be used to investigate environmental change on the local, regional, and global levels by relating variations in rock chemistry to changes in the environment in which the sediment was deposited. An extreme example of this type of investigation might be the discovery of strata rich in iridium near the boundary between the Cretaceous and Tertiary Systems globally. The high concentration of iridium, which is generally rare in the Earth's crust, is indicative of a large delivery of extraterrestrial material, presumably from a large asteroid impactor during this time. A more prosaic example of chemostratigraphic reconstruction of past conditions might be the use of the Carbon-13/Carbon-12 ratio over geologic time as a proxy for changes in carbon cycle processes at different stages of biological evolution. Second, regionally or globally correlatable chemostratigraphic signals can be found in rocks whose formation time is well-constrained by radionuclide dating of the strata themselves or by strata easily correlated with them, such as a volcanic suite that interrupts nearby strata. However, many sedimentary rocks are much harder to date, because they lack minerals with high concentrations of radionuclides and cannot be correlated with nearly datable sequences. Yet many of these rocks do possess chemostratigraphic signals. Therefore, the correlation between chemostatigraphic signals in conventionally datable and non-datable sequences has extended greatly our understanding of the history of tectonically quiescent regions and of biological organisms that lived in such regions. Chemostratigraphy also has acted as a check on other sub-fields of stratigraphy such as biostratigraphy and magnetostratigraphy.

One interesting aspect of chemostratigraphy from the perspective of stratigraphic formalism is that there is no generally accepted definition or classification scheme for chemostratigraphic units in the North American Stratigraphic Code. Chemostratigraphers generally speak in the terms of radiogenic dates or chronostratigraphic units when referring to particular periods or events of regional or global importance.

TECTONOSTRATIGRAPHY

In geology, tectonostratigraphy is stratigraphy that refers either to rock sequences in which large-scale layering is caused by the stacking of thrust sheets, or nappes, in areas of thrust tectonics or to the effects of tectonics on lithostratigraphy.

Tectonically Formed Stratigraphy

One example of such a tectonostratigraphy is the Scandinavian Caledonides. Within the entire exposed 1800 km length of this orogenic belt the following sequence is recognised from the base upwards:

- Autochthon

undisturbed foreland of the Baltic plate

- Parautochthon

thrust sheets that have moved only a short distance (up to 10s of km) from their original position

- Lower allochthon

far travelled thrust sheets derived from the Baltic plate passive margin, mainly sediments associated with the break-up of Rodinia

- Middle allochthon

also derived from the margin of the Baltic plate, Proterozoic basement and its psammitic cover

- Upper allochthon

thrust sheets including island arc and ophiolitic sequences

- Uppermost allochthon

thrust sheets containing sediments with fossil assemblages indicating an origin on the margin of the Laurentian plate

This vertically stacked sequence thus represents the passive margins of Baltica and Laurentia and intervening island arcs and back-arc basins telescoped together and emplaced on top of the Baltic Shield, involving hundreds of km of shortening.

Within this overall stratigraphy the individual layers have their own tectonostratigraphy of stacked thrust sheets.

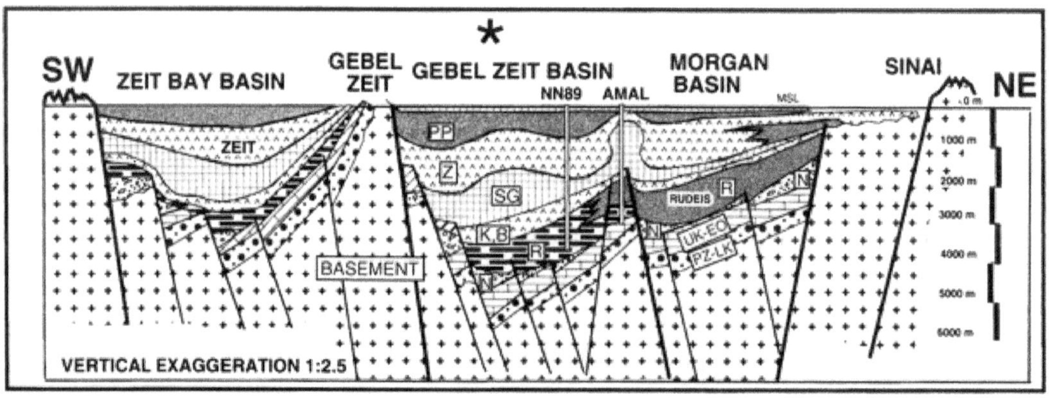

Generalized structural cross-section through the central part of the Gulf of Suez. PZ-LK = Paleozoic to lower Cretaceous Nubia (reservoir rock); UK-EO = Upper Cretaceous to Eocene pre-rift carbonate (source rock); N, R, K, and B = syn- and post-rift Nukhul, Rudeis, Kareem and Belayim formation (sources, reservoirs, seals and overburden); SG = South Gharib salt (seal and overburden); Z=Zeit (seals and overburden); and PP = Plio-Pleistocene (overburden)

Effects of Active Tectonics on Lithostratigraphy

Tectonic events are typically recorded in sediments being deposited at the same time. In the case of a rift, for instance, the sedimentary sequence is normally broken down into three parts:

- The *pre-rift* includes a sequence deposited before the onset of rifting, recognised by the lack of thickness and sedimentary facies changes across the rift faults.
- The *syn-rift* includes a sequence deposited during active rifting, typically showing facies and thickness changes across the active faults, unconformities on the fault footwalls may pass laterally into continuous conformable sequences in the hanging walls.
- The *post-rift* includes a sequence deposited after the rifting has finished, it may still show thickness and facies changes around the rift faults due to the effects of differential compaction and remnant rift topography, particularly in the earliest part of the sequence.

This relatively straightforward nomenclature may become difficult to use, however, in the case of multiphase rifting with the post-rift from one event being the pre-rift to a later event.

CYCLOSTRATIGRAPHY

Cyclostratigraphy is the study of astronomically forced climate cycles within sedimentary successions. Astronomical cycles are variations of the Earth's orbit around the sun due to the gravitational interaction with other masses within the solar system. Due to this cyclicity solar irradiation differs through time on different hemispheres and seasonality is affected. These insolation variations have influence on earth's climate and so on the deposition of sedimentary rocks.

The main orbital cycles are precession with at present main periods of 19 and 23 kyr, obliquity with at present main periods of 41 kyr, and 1.2 Myr, and eccentricity with at present main periods of around 100 kyr, 405 kyr, and 2.4 Myr.

Cyclostratigraphic study of rock records can lead to accurate dating of events in the geological past, to increase understanding of cause and consequences of earth's (climate) history, and to more control on depositional mechanisms of sediments and the acting of sedimentary systems.

The nature of sediments can vary in a cyclic fashion, and these cycles can be displayed in the sedimentary record - here visible in the coloration and resistance of strata.

Cyclic Sediments

Cyclic sediments (also called rhythmic sediments) are sequences of sedimentary rocks that are characterized by repetitive patterns of different rock types (strata) within the sequence. Cyclic sediments can be identified as either autocyclic or allocyclic, and can be hundreds or even thousands of meters thick. The study of sequence stratigraphy was developed from controversies over the causes of cyclic sedimentation.

Processes that Create Cyclic Sediments

Cyclic sedimentation occurs when there is a repetition of a specific series of connected events that affects the environment the sediments are deposited in. Changes in the environment of deposition change the type and amount of sediments that are deposited, resulting in different sedimentary rock types. A series of connected events produces a distinctive and repetitive series of strata that are arranged vertically. At least one rock type, which is regarded as the starting point, must be repeated.

Based on the processes that generate the cyclic deposits, two types of sedimentary cyclic successions — autocycles and allocycles — can be distinguished.

Autocycles

Autocycles are cyclic sediments that are created by processes that only take place within the basin that the sediments are deposited in. Tides and storms are examples of the types of processes that create autocycles. Autocycles show limited stratigraphic continuity.

An example of autocyclic sedimentation on a carbonate platform by Septfontaine M. (1985): Depositional environments and associated foraminifera (lituolids) in the middle liasic carbonate platform of Morocco.- Rev. de Micropal., 28/4, 265-289.

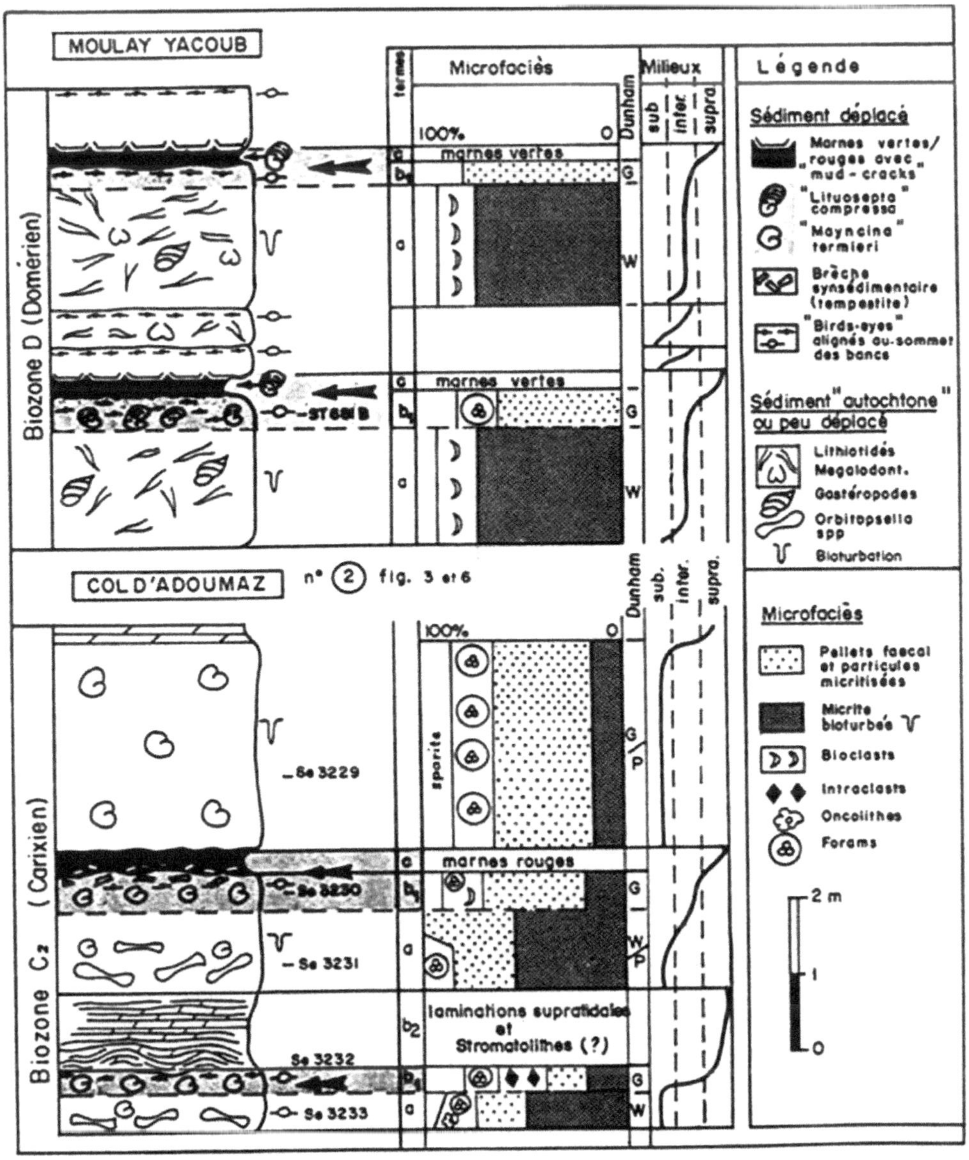

"shallowing upward" sequences from two sections distant of 230 km; note the hurricane (tempestites and tsunami ?)levels with abundant displaced foraminifera on supratidal flat. Middle Liassic, Morocco.

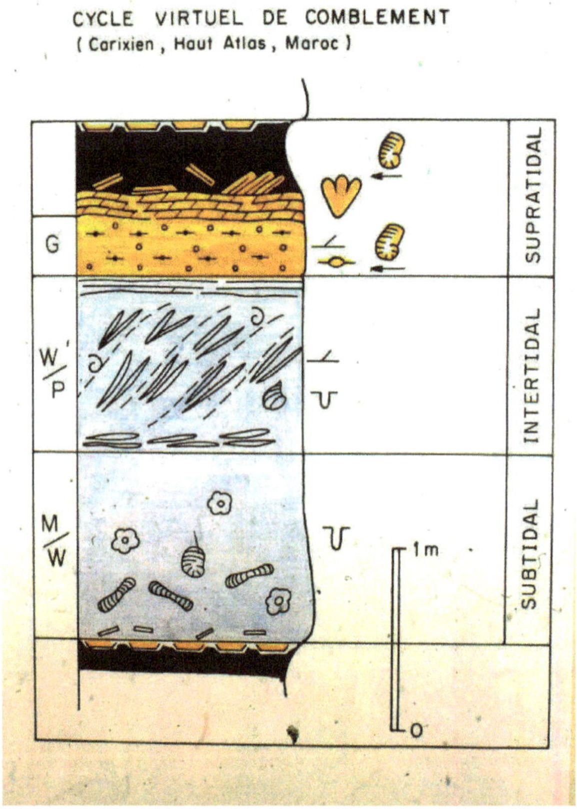

Model of a virtual "shallowing upward" metric sequence observed on carbonate platforms all along the south tethyan margin (about 10.000km) during the Liassic. (Micro)fossils are identical from the Maghreb till Oman and furtheron.

"Shallowing upward" metric first order cycle in the Middle Liassic platform of the High Atlas, Morocco. Algal dolomitized laminations on top.

Recent equivalent of an emersive autocyclic cycle in a lagoonal to supratidal environment. Algal mats in yellow. Tunisia.

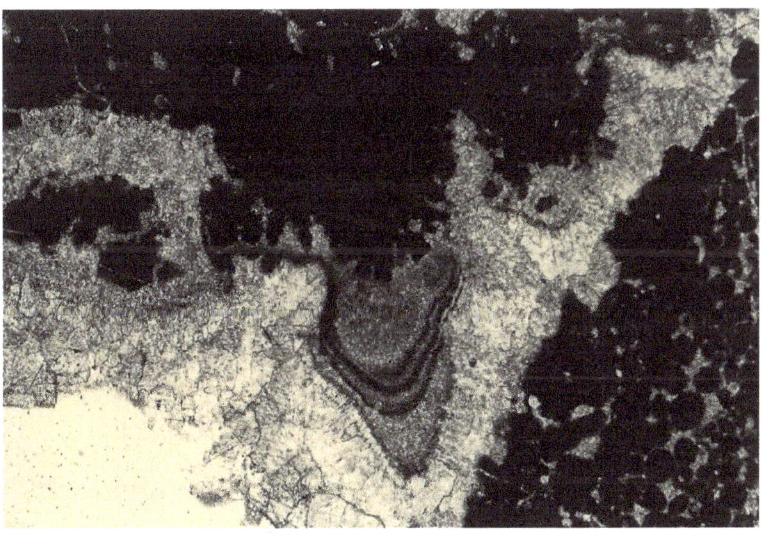

Stalactitic cement from a sample in the supratidal zone in vadose environment (air within the sediment), top of regressive cycle, middle Lias, High Atlas. Thin section, L = 0,5mm.

Hurricane breccia cemented on the surface of a bed, top of a “shallowing upward” cycle. Middle Lias, High Atlas.

Top of a metric regressive cycle with desiccations, surface of bed dolomitised. Middle liassic of the High Atlas, Morocco.

Washover, ammonites and belemnites (by storms) on top of a regressive cycle, supratidal environment. Middle liassic of the High Atlas, Morocco.

Giant dinosaur tracks, in a muddy sediment on top of a metric regressive cycle. Middle liassic, High Atlas, Morocco.

Metric to hectometric regressive cycles (related to an acceleration of the rate of subsidence); South of the High Atlas, Morocco.

Seasonal changes in weather can create cyclic sediments in the form of alternating bands of clay and silt (also known as varves). For example, in a glacial region where sediments are deposited in a lake, coarse sediments that are trapped in ice are released when the ice melts in the summer. This creates paler, coarser silt bands in the lake deposits. In winter, melting is at a minimum, meaning that only fine material is supplied to the lake, causing thin clay layers. Because the cycles are limited to the depositional basin, the lateral extent of the resultant strata are limited. The time period over which autocycles form are usually less than the time periods of allocyclic deposits.

Allocycles

Allocycles are cycles of sediments caused by processes that also occur outside the depositional basin. Sea level fluctuations, climate changes and tectonic activity are examples of these kinds of processes. Allocyclic successions can extend over great distances.

"Shallowing upward" cycles in the lagoonal Liassic platform of the Musandam Peninsula(Oman). Possible allocyclic origin.

Possible allocyclic metric cycles, arranged in decametric sequences, limited by yellow algal, dolomitized levels, below depositional hiatus.

"Shallowing upward" cycles in the Middle Jurassic (Saghtan form.) of the jbel Laghdar Range (Oman).

Changes in sea level can create cyclic sediments of limestones, shales, coals and seat earths. For these cycles to have been created, the environment at the site of deposition must have changed radically from marine to deltaic, then lagoonal and then continental environments. One cause of sea level change is the advance or retreat of continental glaciers caused by climate change. Tectonic movements can affect the environment of deposition by changing water depth. Allocyclic cycles can extend over great distances and are not limited to a single depositional basin. Metric sedimentary carbonate cycles could be related to an astronomical (Milankovitch) precession influence (through thermic dilatation of oceans ?) on an estimated 20.000 years scale, according to numerous authors interpretations. But these beds are of no use in correlation, even at hectometric to kilometric scale, and certainly not a "high resolution" tool for stratigraphy without a severe biostratigraphic control !

Problems with the Study of Cyclic Sediments

The debate about the causes of cyclic sedimentation has been contentious in the past, and it remains unresolved. Sequence stratigraphy, the study of sea level change through the examination of sedimentary deposits, was developed from the centuries-old controversy over the origin of cyclic sedimentation and the relative importance of eustatic and tectonic factors on sea level change.

Another problem with the study of cyclic sediments is that different researchers have different criteria with which they identify cycles and the surfaces that separate the sedimentary layers within the cycles. There is also not a consistent terminology and classification scheme to describe the nature of the cycles seen in the stratigraphic record. This is mainly because absolute age dating is not precise enough at present.

ARCHAEOLOGICAL STRATIGRAPHY

In the field of archaeology, soil stratigraphy is used to better understand the processes that form and protect archaeological sites. Since the law of superposition holds true, it can help date finds or features from each context; these finds and features can be placed in sequence and the dates interpolated. Phases of activity can also often be seen through stratigraphy, especially when a trench or feature is viewed in section (profile). Because pits and other features can be dug down into earlier levels, not all material at the same absolute depth is necessarily of the same age; close attention has to be paid to the archeological layers. The Harris-matrix is a tool to depict complex stratigraphic relations when they are found, for example, in the context of urban archaeology.

Principles or "Laws"

The principle of superposition establishes that within a series of layers and interfacial features, as originally created, the upper units of stratification are younger and the lower are older, for each must have been deposited on, or created by the removal of, a pre-existing mass of archaeological stratification.

The principle of original horizontality states that any archaeological layer deposited in an unconsolidated form will tend towards a horizontal deposition. Strata which are found with tilted surfaces were so originally deposited, or lie in conformity with the contours of a pre-existing basin of deposition. The principle of lateral continuity states that any archaeological deposit, as originally laid down, will be bounded by the edge of the basin of deposition, or will thin down to a feather edge. Therefore, if any edge of the deposit is exposed in a vertical plane view, a part of its original extent must have been removed by excavation or erosion: its continuity must be sought, or its absence explained.

The principle of stratigraphic succession states that any given unit of archaeological stratification exists within the stratigraphic sequence from its position between the undermost of all higher units and the uppermost of all lower units and with which it has a physical contact.

Combining Stratigraphic Contexts for Interpretation

Understanding a site in modern archaeology is a process of grouping single contexts together in ever larger groups by virtue of their relationships. The terminology of these larger clusters varies depending on the practitioner, but the terms interface, sub-group, and group are common. An example of a sub-group could be the three contexts that make up a burial; the grave cut, the body, and the back-filled earth on top of the body. Sub-groups can then be clustered together with other sub-groups by virtue of their stratigraphic relationship to form groups, which in turn form "phases." A sub-group burial could cluster with other sub-group burials to form a cemetery, which in turn could be grouped with a building, such as a church, to produce a "phase". Phase implies a nearly contemporaneous Archaeological horizon, representing "what you would see if you went back to time X".The production of phase interpretations is the first goal of stratigraphic interpretation and excavation.

Stratigraphic Dating

Archaeologists investigating a site may wish to date the activity rather than artifacts on site by dating the individual contexts which represents events. Some degree of dating objects by their position in the sequence can be made with known datable elements of the archaeological record or other assumed datable contexts deduced by a regressive form of relative dating which in turn can fix events represented by contexts to some range in time. For example, the date of formation of a context which is totally sealed between two datable layers will fall between the dates of the two layers sealing it. However the date of contexts often fall in a range of possibilities so using them to date others is not a straightforward process.

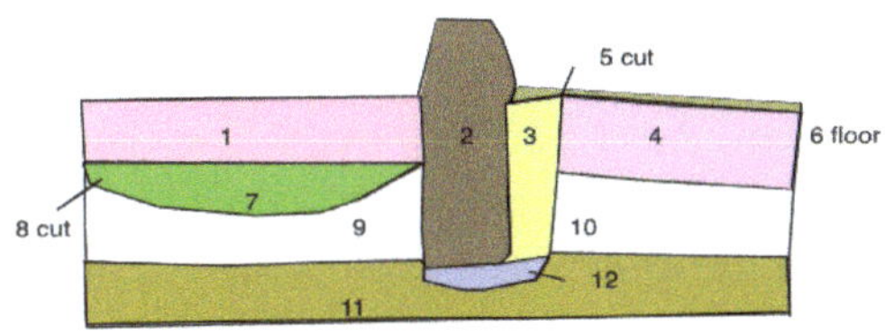

A

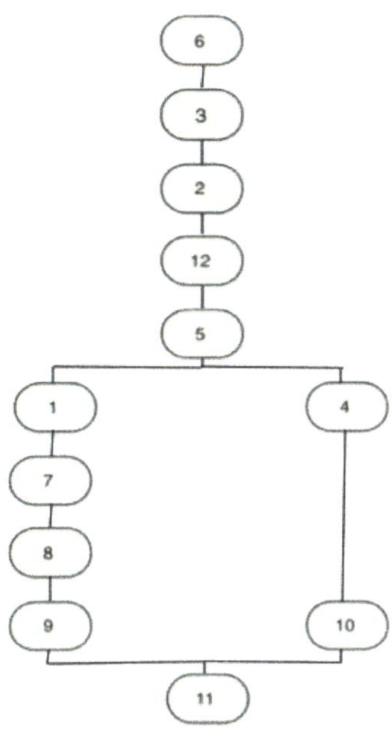

B

Take the hypothetical section *fig A*. Here we can see 12 contexts, each numbered with a unique context number and whose sequence is represented in the Harris matrix in *fig B*.

- A horizontal layer
- Masonry wall remnant
- Backfill of the wall construction trench (sometimes called construction cut)
- A horizontal layer, probably the same as 1
- Construction cut for wall 2
- A clay floor abutting wall 2
- Fill of shallow cut 8
- Shallow pit cut
- A horizontal layer
- A horizontal layer, probably the same as 9
- Natural sterile ground formed before human occupation of the site
- Trample in the base of cut 5 formed by workmen's boots constructing the structure wall 2 and floor 6 is associated with.

If we know the date of context 1 and context 9 we can deduce that context 7, the backfilling of pit 8, occurred sometime after the date for 9 but before the date for 1, and if we recover an assemblage of artifacts from context 7 that occur nowhere else in the sequence, we have isolated them with a reasonable degree of certainty to a discrete range of time. In this instance we can now use the date we have for finds in context 7 to date other sites and sequences. In practice a huge amount of cross referencing with other recorded sequences is required to produce dating series from stratigraphic relationships such as the work in seriation.

Residual and Intrusive Finds

One issue in using stratigraphic relationships is that the date of artifacts in a context does not represent the date of the context, but just the earliest date the context could be. If one looks at the sequence in *fig A*, one may find that the cut for the construction of wall 2, context 5, has cut through layers 9 and 10, and in doing so has introduced the possibility that artifacts from layers 9 and 10 may be redeposited higher up the sequence in the context representing the backfill of the construction cut, context 3. These artifacts are referred to as "residual" or "residual finds". It is crucial that dating a context is based on the latest dating evidence drawn from the context. We can also see that if the fill of cut 5 – the wall 2, backfill 3 and trample 12 — are not removed entirely during excavation because of "undercutting", non-residual artifacts from these later "higher" contexts 2, 3 and 12 could contaminate the excavation of earlier contexts such as 9 and 10 and give false dating information. These artifacts may be termed *intrusive finds*.

REFERENCES

1. Davies G.L.H. (2007). Whatever is Under the Earth the Geological Society of London 1807-2007. London: Geological Society. p. 78. ISBN 9781862392144.
2. SEPM Strata, Society for Sedimentary Geology, Terminology=hiatus
3. Martinsen, O. J. et al. (1999) "Cenozoic development of the Norwegian margin 60–64N: sequences and sedimentary response to variable basin physiography and tectonic setting" pp. 293–304 In Fleet, A. J. and Boldy, S. A. R. (editors) (1999) Petroleum Geology of Northwest Europe Geological Society, London, page 295, ISBN 978-1-86239-039-3
4. Kearey, Philip (2001). Dictionary of Geology (2nd ed.) London, New York, etc.: Penguin Reference, London, p. 123. ISBN 978-0-14-051494-0.
5. Chapman, Richard E. (1983) Petroleum Geology Elsevier Scientific, Amsterdam, page 33, ISBN 978-0-444-42165-4
6. Gluyas, J. & Swarbrick, R. (2004) Petroleum Geoscience. Publ. Blackwell Publishing. pp. 80-82
7. Gehling, James; Jensen, Sören; Droser, Mary; Myrow, Paul; Narbonne, Guy (March 2001). "Burrowing below the basal Cambrian GSSP, Fortune Head, Newfoundland". Geological Magazine. 138 (2): 213–218. doi:10.1017/S001675680100509X. 1.
8. An updated Telychian (Late Llandovery, Silurian) conodont zonation based on Baltic faunas. Peep Männik, Lethaia, Volume 40, Issue 1, pages 45–60, March 2007, doi:10.1111/j.1502-3931.2006.00005.x
9. Juia A. Jackson, ed. (1997). Glossary of Geology (4th ed.). Alexandria, VI: American Geological Institute. ISBN 0-922152-34-9.
10. Fastovsly, David E.; Weishampel, David B. (1996). The Evolution and Extinction of the Dinosaurs. Cambridge [u.a.]: Cambridge Univ. Press. ISBN 0-521-44496-9.
11. Smith, J. B.; Lamanna, M.C.; Lacovara, K.J.; Dodson, P. Jr.; Poole, J.C.; Giegengack, R. (1 June 2001). "A Giant Sauropod Dinosaur from an Upper Cretaceous Mangrove Deposit in Egypt". Science. 292 (5522): 1704–1706. doi:10.1126/science.1060561. ISSN 1095-9203. PMID 11387472. Retrieved 2 February 2013.
12. International Commission on Stratigraphy. "Chronostratigraphic Units." International Stratigraphic Guide. Accessed 14-DEC-2009.
13. Prothero, D.R.; Schwab, F. (2014). "Section IV: Stratigraphy. Ch. 17: Geophysical and Chemostratigraphic Correlation". Sedimentary Geology (3rd ed.). W.H. Freeman. ISBN 978-1-4292-3155-8.
14. Berger, W.H.; Vincent, E. (1981). "Chemostratigraphy and biostratigraphic correlation: exercises in systematic stratigraphy". Oceanol. Acta: 115–127.
15. Renard, M.; Corbin, J.C.; Daux, V.; Emmanuel, L.; Baudin, F.; Tamburini, F. (2008). "Ch. 3: Chemostratigraphy". In Rey, Jacques; Galeotti, Simone. Stratigraphy: Terminology and Practice. Editions OPHRYS. pp. 41–52. ISBN 978-2-7108-0910-4.

16. Roberts,D. & Gee,D. 1985. An introduction to the structure of the Scandinavian Caledonides. In Gee, D. G., and Sturt, B. A., eds. The Caledonide Orogen - Scandinavia and related areas.John Wiley and Sons, Chichester, 55-68.
17. Jackson, C A L ,Gawthorpe, R L, Leppard, C W , Sharp, I R 2006. Rift-initiation development of normal fault blocks: insights from the Hammam Faraun fault block, Suez Rift, Egypt. Journal of the Geological Society, 163, 165-183.
18. Andre Strasser, Frederik Hilgen, Philip H. Heckel. "Cyclostratigraphy - from orbital cycles to geologic time scale". 2008. http://www.cprm.gov.br/33IGC/1312131.html
19. Hinnov L.A. & Ogg J.G. (2007). "Cyclostratigraphy and the Astronomical Time Scale" (PDF). Stratigraphy. 4 (2-3): 239–251.
20. Isabelle Cojan et Maurice Renard, Sédimentologie, 2e éd., Dunod, 2006, ISBN 2-10-049623-9 (French)
21. Graham P. Weedon, Time-Series Analysis And Cyclostratigraphy: Examining Stratigraphic Records of Environmental Cycles, Cambridge University Press, 2005 ISBN 0-521-01983-4 (English)
22. André Berger, Le climat de la Terre — Un passé pour quel avenir ?, De Boeck université, 1992, ISBN 2-8041-1497-X (French)
23. Brigitte Van Vliet-Lanoë, La planète des glaces — Histoire et environnements de notre ère glaciaire, Vuibert, 2005, ISBN 2-7117-5377-8 (French)
24. Patrick de Wever, Loïc Labrousse, Daniel Raymond, André Schaaf, La mesure du temps dans lɒhistoire de la Terre, Société géologique de France, édité par Vuibert, 2005, ISBN 2-7117-5393-X (French)

Index

R

S

T